W9-CMQ-448

STAR TREK™

THE SCIENCE OF STAR TREK FROM TRICORDERS TO WARP DRIVE

First published in 2017 by Voyageur Press, an imprint of The Quarto Group, 401 Second Avenue North, Suite 310, Minneapolis, MN 55401 USA. T (612) 344-8100 F (612) 344-8692 www.quartoknows.com

Voyageur Press titles are also available at discount for retail, wholesale, promotional, and bulk purchase. For details, contact the Special Sales Manager by email at specialsales@quarto.com or by mail at The Quarto Group, Attn: Special Sales Manager, 401 Second Avenue North, Suite 310, Minneapolis, MN 55401 USA.

10 9 8 7 6 5 4 3

ISBN: 978-0-7603-5263-2

Library of Congress Cataloging-in-Publication Data

Names: Siegel, Ethan, 1978-
Title: Treknology : the science of Star Trek, from tricorders to warp drive / Ethan Siegel, PhD.
Other titles: Science of Star Trek, from tricorders to warp drive
Description: Minneapolis, MN : Voyageur Press, 2017. | Includes index.
Identifiers: LCCN 2017013887 | ISBN 9780760352632 (hc)
Subjects: LCSH: Science--Forecasting--Popular works. | Technological forecasting--Popular works. | Star trek (Television program)
Classification: LCC Q162 .S4854 2017 | DDC 601/.12--dc23
LC record available at https://lccn.loc.gov/2017013887

Acquiring Editor: Madeleine Vasaly
Project Manager: Alyssa Lochner
Art Director: James Kegley
Cover & Page Designer: Amelia LeBarron
Layout: Kim Winscher

Printed in China

CONTENTS

INTRODUCTION: THE FINAL FRONTIER

When you were young, what did you dream of when you looked to the future? Did it have anything to do with the exploration of the unknown? Of pushing back the frontiers and limits of our knowledge? Of discovering new worlds, new forms of life, or even entirely new civilizations? In some ways, the most curious of our human instincts is the pioneering spirit: to go where no one has gone before. Immediately following the launch of *Sputnik 1* in October of 1957, the United States scrambled to put a satellite into orbit; in January of 1958, *Explorer 1* was launched. Another two months after that, the White House made public a report from the President's Science Advisory Committee that contained the following sentiment:

> *The first of these factors is the compelling urge of man to explore and to discover, the thrust of curiosity that leads men to try to go where no one has gone before. Most of the surface of the earth has now been explored and men now turn on the exploration of outer space as their next objective.*

Over the next few years, NASA was formed, the Mercury and Gemini programs propelled the United States into the lead in the Space Race, and the Apollo program promised to take humanity to the moon. Our eyes were turned skyward as never before: with hope for the real, tangible possibility that we might become a spacefaring civilization.

Space was a natural place to put our dreams—as well as our fears—about the future. Alien invasion stories dominated science fiction in the twentieth century. *Lost in Space* focused on the consequences of human traits like laziness, selfishness, and nationalism during the Cold War. But when the original *Star Trek* series first aired, it presented a far different view of the future than any that had come before. Rather than humanity fighting an evil alien presence or continuing our own petty, earthly squabbles in a different setting, we had become a peaceful race. We had joined together in an alliance for exploration, knowledge sharing, and mutual assistance with the races of other worlds in a United

Federation of Planets. And instead of highlighting the apocalyptic fears many of us had of new technologies, particularly with the threat of nuclear war hanging over the world's head, *Star Trek* brought us a universe in which advanced technologies were used to bring unequivocal good to the galaxy.

The futuristic dream of what an increased investment and valuing of science and technology might bring to the world is one that has not only remained with us, but been borne out by the past five decades. Computers, barely powerful enough to calculate astronomical trajectories and so large that one took up an entire room at the inception of *Star Trek*, were envisioned to carry on natural-language conversations, accept and execute verbal commands, and communicate and network with other vessels and starbases, even light-years away. Handheld devices, ranging from phasers to communicators to electronic clipboards, envisioned a rise in widespread technology that we now carry around as ubiquitously as our keys. Medical devices could scan, diagnose, and even treat the sick or injured without ever touching their skin, portending many of the advances we see today. And civilian technologies like

The AS-201 was the first test flight of the Apollo Command/Service Module and the Saturn IB launch vehicle, launched February 26, 1966, mere months before *Star Trek*'s September 8 premiere.

replicators and sliding doors have sprung to life with novel developments like 3D printing in ways that would seem like magic to someone dropped from the 1960s into the present day.

Star Trek wasn't afraid to dream big, either. Technologies that seemed to defy physics—warp drives that propelled starships faster than light, transporters that would dematerialize you and rematerialize you in a different location, cloaking devices that could render even a starship completely invisible—were ubiquitous in *Star Trek*'s vision of what the future would hold for us centuries from now. When *Star Trek: The Next Generation* began airing in the 1980s, the envisioned future had grown even more advanced than it had been in the original series, with more ambitious and more sleekly designed technologies making their appearance. Remarkably, we've made tremendous progress toward a great many of these. Much of what once seemed to be completely impossible is now inching closer to reality, with everything from artificial eyes to holograms to androids coming into use across the world.

But part of the reason *Star Trek* has had such a lasting cultural impact is not just the futuristic technologies envisioned by the show, but the ethical and moral questions it brought up as well. Those questions are no less omnipresent in our world today, and are often intimately tied to the technologies themselves. As artificial intelligence efforts continue to improve at exponential rates, at what point do we begin granting rights to machines? As information becomes freer and more widespread, what does that mean for our rights to privacy, or for our mistakes to be forgiven and forgotten? If we can alter our unborn children's genetic makeup—or even our own, retroactively—what are the bioethical implications of doing so? If we can perform medical procedures or install implants to restore or enhance patients' biological functions, should we always do so, even if there are security risks to someone's own body being hacked? And if you could successfully beam an individual from one location to another, how certain can you be that the transported human who arrives at the destination is actually the same person that you transported, and not an identical copy who replaces the now-deceased original?

More than fifty years after *Star Trek*'s inception, its legacy continues to capture our collective imaginations. Our desire to push the boundaries of what we're capable of knowing, inventing, and accomplishing is placed front and center in *Star Trek*, right alongside the importance of remaining true to the very things that make us human. Many of the dream technologies envisioned decades ago have already come to fruition; many others are well on their way to becoming reality. As we look at the real-life science and technology behind the greatest advances anticipated by *Star Trek*, it's worth remembering that the greatest legacy of the show is its message of hope. The future can be brighter and better than our past or present has ever been. It's our continuing mission to make it so.

No matter where in the galaxy we are, even on a voyage that takes us a long way from home, a combination of advanced technology and our spirit of inventiveness and ingenuity gives us every reason to hope that we can accomplish even the most seemingly impossible goals.

STARSHIP

TECHNOLOGY

"Space. The final frontier. These are the voyages of the *Starship Enterprise*."

Since humanity first turned its eyes skyward, it's been a dream of our species not only to discover what's out there beyond our homeworld, but to explore it. Over the past few hundred years, we've come to discover that there are hundreds of billions of stars in our galaxy alone, most of which contain planets, some of which are rocky, at the right distance for liquid water on their surface, and possibly teeming with life. The rest of the universe awaits us, full of perhaps even more potential for what we might discover than the most imaginative *Star Trek* writers and creators ever envisioned. But getting there relies on a lot more than just imagination; we need starship technology capable of transporting human beings safely across great interstellar distances if we ever want to seek out new life and new civilizations.

In order to safely transport a crew, equipment, and a starship from one region of space to another, a variety of technologies are required: short-range propulsion for precision movements around a planetary system; long-range engines for traveling near, at, or faster than the speed of light; a safe way of containing the enormously powerful fuel required for the journey; plus a means of transporting crews and equipment from the ship down to the surface. Without the development of these key technologies, traveling to and exploring a star system beyond our own would be an impossibility for a single human lifetime.

Yet *Star Trek* was never about succumbing to the present limits of human technology, but insisted on extending them to transform the impossible into the possible. Laws of physics could be cleverly subverted when they couldn't be broken, and human ingenuity enabled a vision of the future in which technology has brought about a virtual paradise. While human nature still ensures that there will be vices like greed, callousness, self-interest, and violence, *Star Trek*'s future is also one full of generosity, altruism, kindness, and peace. It is a future of exploration and collaboration, not of conquest and competition. It is a future that still seems far away today—but possible. More than fifty years after *Star Trek* premiered, our dreams of traveling to the stars are closer

The Alpha Quadrant and Gamma Quadrant are connected by a stable wormhole, an example of a topological defect, in *Star Trek: Deep Space Nine*.

WARP DRIVE

"Helm, warp 1, engage!" The warp nacelles power up, and the speed of light is no match for the *Enterprise*. Whether the need is pressing, such as escaping a supernova or fleeing from a more powerful opponent, or less urgent travel to a distant destination, warp drive provides a way for space travelers to exceed the speed of light by many orders of magnitude. This enables crews to cut the travel time between the stars from years down to days or even hours.

Zefram Cochrane's first warp flight in 2063 led to direct contact with the Vulcans, and continued advances meant that higher warp factors and reduced travel times—with warp 1 just exceeding the light barrier and warp 10 corresponding to infinite speeds—became accessible over time. Captain Archer's *Enterprise* could achieve warp 5.2; Kirk's could safely reach warp 8; Picard's could travel up to warp 9.8 at extreme risk; and Janeway's *Intrepid*-class *Voyager* could reach warp 9.975. This last mark tells us—for a seventy-five-year journey at a distance of 70,000 light-years from home—that *Voyager* could travel at nearly one thousand times the speed of light on an ongoing basis.

Even before the inception of *Star Trek*, finding a way to defeat the speed of light seemed a necessity for human space exploration. Given that the nearest star to our own sun is more than 4 light-years away, traveling anywhere would mean multiple years passing back on Earth, even if a ship itself took advantage of relativity to shorten the journey for the crew. According to Einstein's theory of relativity, when you travel close to the speed of light, the distances in your direction of motion appear shorter (length contraction) and the rate at which time passes appears slower (time dilation), two of the most counterintuitive and yet well-studied and confirmed consequences of special relativity. If this were all there is to traveling through the universe, then crew members traveling at near-light speeds would remain relatively young, while years would pass both at the origin and destination star systems. Interstellar travel would be a generational venture for all but the absolute nearest stars.

Yet general relativity offers a possible escape from this constraint—through the malleability of the fabric of spacetime itself. We might be unable to travel through space itself at speeds greater than 299,792,458 meters per second, but if we can lessen the actual distances between two locations (or events), then we could travel there quickly not only from the crew's perspective, but from the perspective of observers at both the source and the destination. There are two very different solutions that Einstein's theory offers to those wishing to travel great distances without the space-and-time constraints imposed by conventional travel through interstellar (or intergalactic) space: by connecting two otherwise disconnected points through a topological defect or by distorting (shortening) the space along a starship's direction of motion.

The idea of a wormhole is a realization of connecting two distant points through a cosmic "shortcut," or a bridge between locations in the fabric of space. Physically, this requires extraordinarily strong curvatures to space itself, which can only be realistically achieved through the creation of black holes or other cosmic singularities. While the idea of connecting two black holes (or a black hole and a white hole) by a spacelike bridge is theoretically possible in general relativity—known as an Einstein-Rosen bridge after the paper the two scientists cowrote in 1935—it is unstable and untraversable unless additional physics beyond what's currently accepted are invoked. The wormhole requires either an additional scalar field (modifying and extending the gravitational laws of physics), some form of exotic (for example, negative-energy) matter, or the existence of accessible extra

A warp drive, or an engine capable of directed faster-than-light travel, would be required to traverse any set of chosen interstellar distances in a reasonable amount of time.

dimensions. Even if these entities did exist, wormholes would simply exist between two fixed points in space (like the Bajoran wormhole on *Star Trek: Deep Space Nine*) and would not allow faster-than-light travel between any two locations at will.

But the other option, distorting space along the direction of motion (in front) of the spacecraft, is exactly the idea that *Star Trek*'s warp field purports. Put forth in a hand-waving sort of fashion by science fiction writers in the 1960s, warp drive could effectively speed up a journey across the stars arbitrarily, limited only by how dramatically one could shorten the space in front of you. But this was shown to be a real possibility within our own universe in 1994, when Mexican theoretical physicist Miguel Alcubierre successfully reverse-engineered a solution within general relativity that created exactly this type of spacetime. By shortening the space in front of you and lengthening the space behind you by an equal-and-opposite amount, while creating a stable "bubble" of space inside for your starship to reside in, the physics behind travel at warp speed—now known as the Alcubierre drive—went from science fiction to plausible science.

There are a great number of obstacles to be overcome, however—both hypothetical and pragmatic—in order for the Alcubierre drive to become a reality. For one, the most conservative estimate for the energies required to deform any nonempty region of space in this fashion equates to at least 20,000 megatons of TNT, or a ton of mass converted into pure energy, according to Einstein's $E = mc^2$. For another, the Alcubierre drive requires the creation of a region of space with an energy that's less than the zero-point energy of space itself, requiring the existence of negative mass (or negative energy) in some form. While that might seem an insurmountable constraint, as only positive masses and energies are known to exist in this universe, a setup similar to the Casimir effect—in which parallel conducting plates can reduce the effective zero-point energy of the space inside—might provide the required energy conditions, a possible solution suggested by Alcubierre himself. Finally, there isn't a known way to either begin traveling at warp speed or to end

A two-dimensional projection of the Alcubierre spacetime, in which space itself is shortened in front of the spacecraft (to the right) and lengthened behind it (to the left).

Contacting another intelligent, spacefaring civilization would mark a realization of the ultimate dream of *Star Trek* for many. Zefram Cochrane, inventor of the warp drive, was the first recorded human to travel faster than light and led to the first official contact with the Vulcans.

your warp-speed travel once it has begun. Clearly, the ability to control your starship requires both! From a pragmatic point of view, unless the tremendous tidal forces surrounding the edge of the warp field are somehow avoided, the ship would be torn apart, causing multiple hull breaches. If you were far enough from the warp field's edge at all times, however, you'd simply travel through space along with the rest of the field at speeds far exceeding the speed of light, experiencing the journey as though you were simply in gravitational free fall.

If such a technology could be harnessed, it would mean great advances on a number of fronts for humanity. For one, we could ship anything—from goods to resources to people—across arbitrarily large distances in correspondingly small amounts of time. Messages about upcoming catastrophes could be delivered before a light signal could ever arrive, and violating our traditional notions of causality would become a routine game (see page 86 for more on subspace communications). But above all, the development of this technology would allow humans to travel across the galaxy, experiencing other stars, other planets, and, if we're lucky, other civilizations. In many aspects, this is the most important advance humanity could strive for in achieving the dreams of *Star Trek*.

But while our dreams may be limitless, our technological progress on this front has been minimal, unfortunately. While a great many scientists and engineers are (understandably!) excited about the prospects of warp drive becoming a reality, the truth of the matter is that even the most advanced and resourceful tests humanity is capable of performing are not only insufficient to create warp drive, they're not yet up to the task of determining whether this mathematical solution is physically plausible in our universe. While there are teams of people at NASA Eagleworks and elsewhere working on this, there have, to date, been no substantive tests that support or refute the possibility of a warp drive. A tremendous leap forward was made in 1994 with the discovery that warp drive was consistent with general relativity, moving this technology from completely speculative into the "theoretically plausible" category. To move forward to the next stage, however, will require technologies and energies beyond the current imaginings of early twenty-first-century physicists. And who knows? If and when we do finally figure it out, we just might wind up going farther and faster than even our vast imaginations have taken us.

TRACTOR BEAMS

"Tractor beam, Captain. Something's grabbed us." Just like that, the ship can no longer control its own motion. Whether it's a combat maneuver, like a Federation vessel being immobilized by a species with greater technology, such as the Borg, or a friendly, gentle tow of equipment or a disabled ship, a tractor beam will take care of you. Far superseding the precision of ropes, cables, or other grappling devices, a *Star Trek* tractor beam takes advantage of a beam of gravitons to attract, repel, or hold another

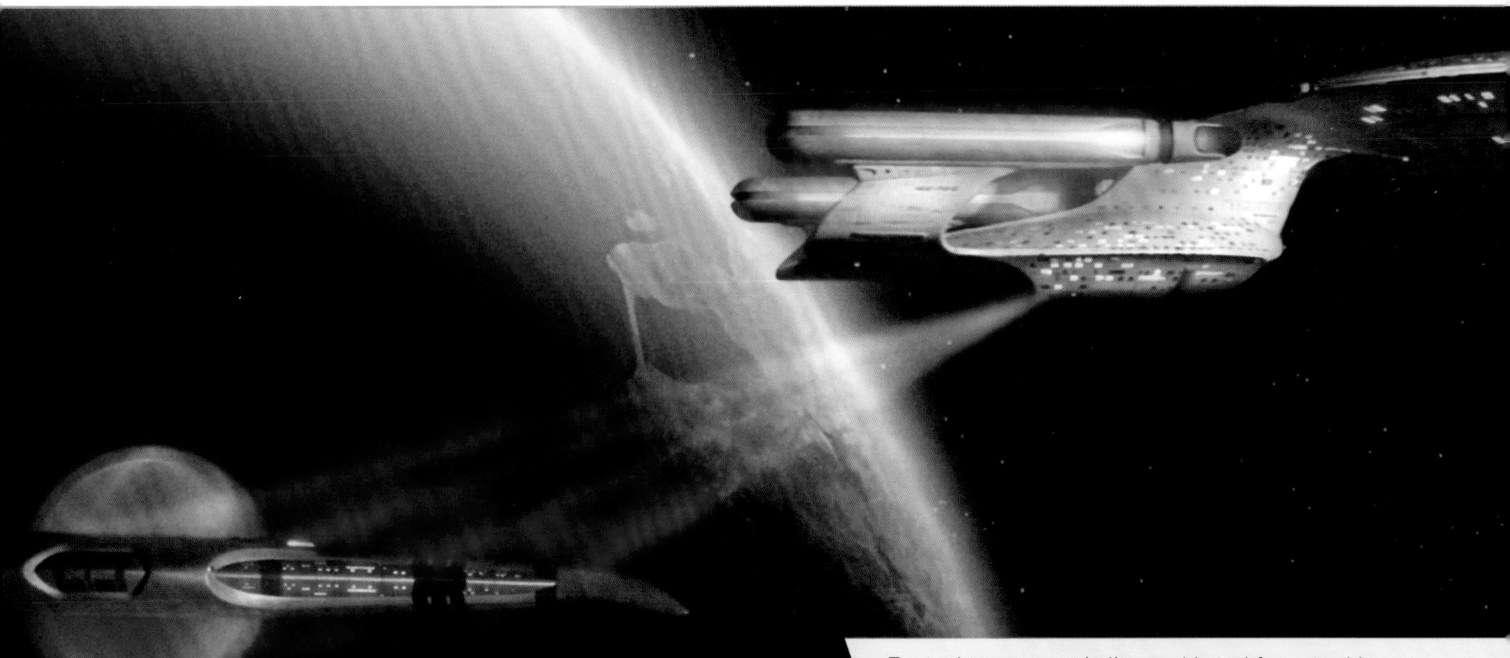

Tractor beams are an indispensable tool for a starship.

The combination of a Borg cube's shield-draining tractor beam along with the cutting beam destroyed a great many Federation starships, including the *U.S.S. Saratoga*. Fortunately, the *Enterprise*-D survived every such encounter.

massive object in place. Perhaps most spectacularly, it can immobilize a starship attempting to use the Picard Maneuver, holding it in place and preventing it from warp-jumping to a strategically advantageous location.

Tractor beams weren't available to Earth-based ships in the mid-2100s, and the *Enterprise* NX-01 was forced to use grappling devices held in place with magnetic locks. But by the time the 2200s came along, both the Federation and many of its adversaries possessed tractor beam technology, capable of towing vessels for prolonged periods of time. While ships like the *Enterprise* NCC-1701 could only do this at sublight speeds, other civilizations (like the First Federation) could tow ships at warp without placing too great a stress on either ship's hull. By the 2300s, Starfleet tractor beams could be used to safely bring vessels out of warp.

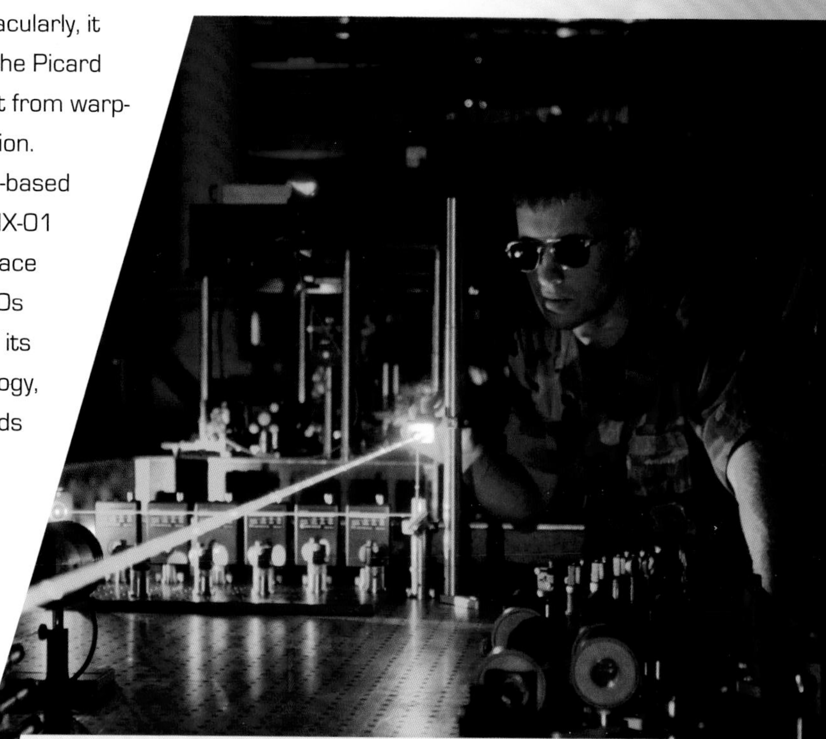

While electromagnetic lasers have been created and utilized since 1960, no progress has been made on a gravitational laser (gaser)—which would be a precursor of a true tractor beam—since the theoretical groundwork was laid more than fifty years ago.

The Borg, however, possessed the most advanced tractor beam of all: one that could grab onto a ship irrespective of its shields. The beam could then drain the shields, rendering the ship helpless in the face of the deadly Borg cutting beam.

While tractor beams would no doubt be a tremendous boon to modern life, revolutionizing everything from freight and human transport to precision manufacturing and assembly, science has a very long way to go. Not only have graviton beams not yet been developed, but gravitons themselves may not even physically exist. The existence of the graviton is predicated on the idea that gravity is an inherently quantum force—that gravitational waves are made of particles, called gravitons, the same

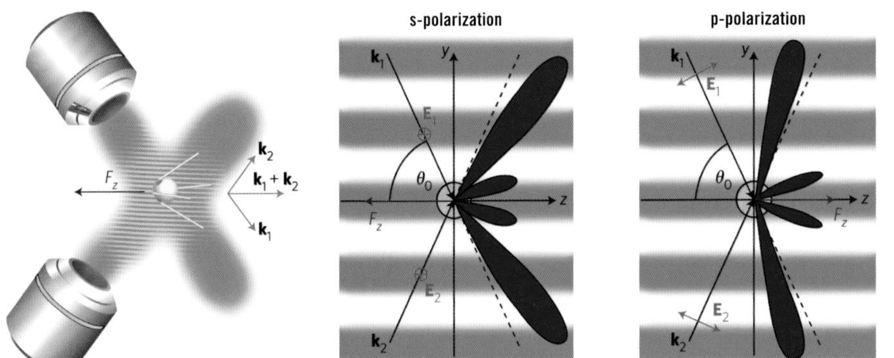

The first working demonstration of an optical "tractor beam" not only pulled particles against the direction of the photon flow, but was able to control their two-dimensional motion and could sort them along the direction of motion.

way that light waves are made of photons—which is something that no experiment, observation, or measurement has yet established. Assuming that gravitons are the force-carrying particles for the gravitational interaction doesn't get you much further: the only substantial thing a spaceship could do to emit gravitons would be to have mass, which doesn't really help. Despite many attempts beginning in the twentieth century to create a beam of gravity-like impulses, no finding has ever been substantiated, reproduced, or published in a peer-reviewed journal. By 2010, the closest anyone had come was a theoretical 1964 paper showing that induced, resonant emissions of gravitons were possible and would create a gravitational version of a laser.

However, the 2010s brought a wave of developments that may lead to a different method for a working tractor beam: one rooted in laser physics. Initial studies have focused on a laser beam shaped like a cylinder, with a hollow, dark core inside. As this beam heats the air around a target particle, the increased temperature causes the particle to be held in place, while the hollow core keeps it cool. This phenomenon is known as photophoresis and was identified in the first half of the twentieth century, but the idea of applying it to create a tractor beam was entirely novel. The critical shortcoming of this device when comparing it to a *Star Trek*–style tractor beam is that it cannot work in the vacuum of space, but it may yet open the door to technologies that will. Sonic tractor beams have also been developed, but they suffer the same limitations, making them unworkable in space.

Normally, if a particle is hit with any light source, it is "pushed" forward—that is, made to move in the same direction of motion as the photon that collided with it. But in 2011, a team of scientists showed it was also possible to "pull" a particle *toward* a light source thanks to a combination of resonant laser light and a magnetic gradient trap. These so-called "optical tweezer" mechanisms cause particles trapped in a beam of light to follow the light's intensity gradient, meaning that particles always reach a particular location where they'll be stably pinned, which of course is exactly what you'd want a tractor beam to do. In 2013, a team based at the University of Saint Andrews came the closest of

As seen in the shuttle bays of the *Enterprise*-D, light-emitting tractor beams could help an incoming shuttlecraft land gently inside the starship.

anyone to developing a working tractor beam: they proved they could use a forward-firing laser to collimate and attract microscopic particles at will. By using multiple lasers simultaneously and by carefully controlling their rotational polarizations, orienting the magnetic fields created by the light in a coherent fashion, they experimentally demonstrated a working tractor beam in a vacuum for the first time. Further studies are ongoing and are anticipated to build upon this incredible development.

It's amazing that for more than forty years after *Star Trek* first brought the tractor beam into the realm of public consciousness, practically no scientific progress was made—yet the willingness to replace the original idea of a graviton-based tractor beam with a laser-based one has brought with it tremendous advances. Not only can individual particles be confined and attracted or repelled until they're at an ideal distance in a medium like air, but massive particles can be attracted toward a destination even in a vacuum. *Star Trek* even arguably anticipated this, as all instances of the tractor beam displayed on the show had not only a visible optical beam but a luminous origin.

Currently, however, real-life tractor beams are effectively limited to trapping and moving microscopic particles. Because being struck with a laser necessarily involves a transfer of energy, any macroscopic object would heat up catastrophically before it was successfully tractored. The current setup might prove to be effective for single cells, but it would destroy a spaceship if it were boosted to the requisite energies. This, too, may have been anticipated by *Star Trek*, since modified tractor beams were used as effective weapons.

Nevertheless, the concept has been validated, and a small-scale working prototype has been developed. Tractor beams are already being considered for microscopic applications. Perhaps macroscopic applications are next on the horizon, if scientists can discover a way to protect the target object from the laser's heat or prevent that heat from being transferred to the object in the first place. It may not look exactly like *Star Trek* imagined it yet, but tractor beam technology has finally been demonstrated to be plausible. The next steps are a job for the next generation of scientists and engineers.

By the twenty-fourth century, tractor beams were so powerful that they could even, with appropriate modifications, be used to move objects as massive as fragments of a star's core.

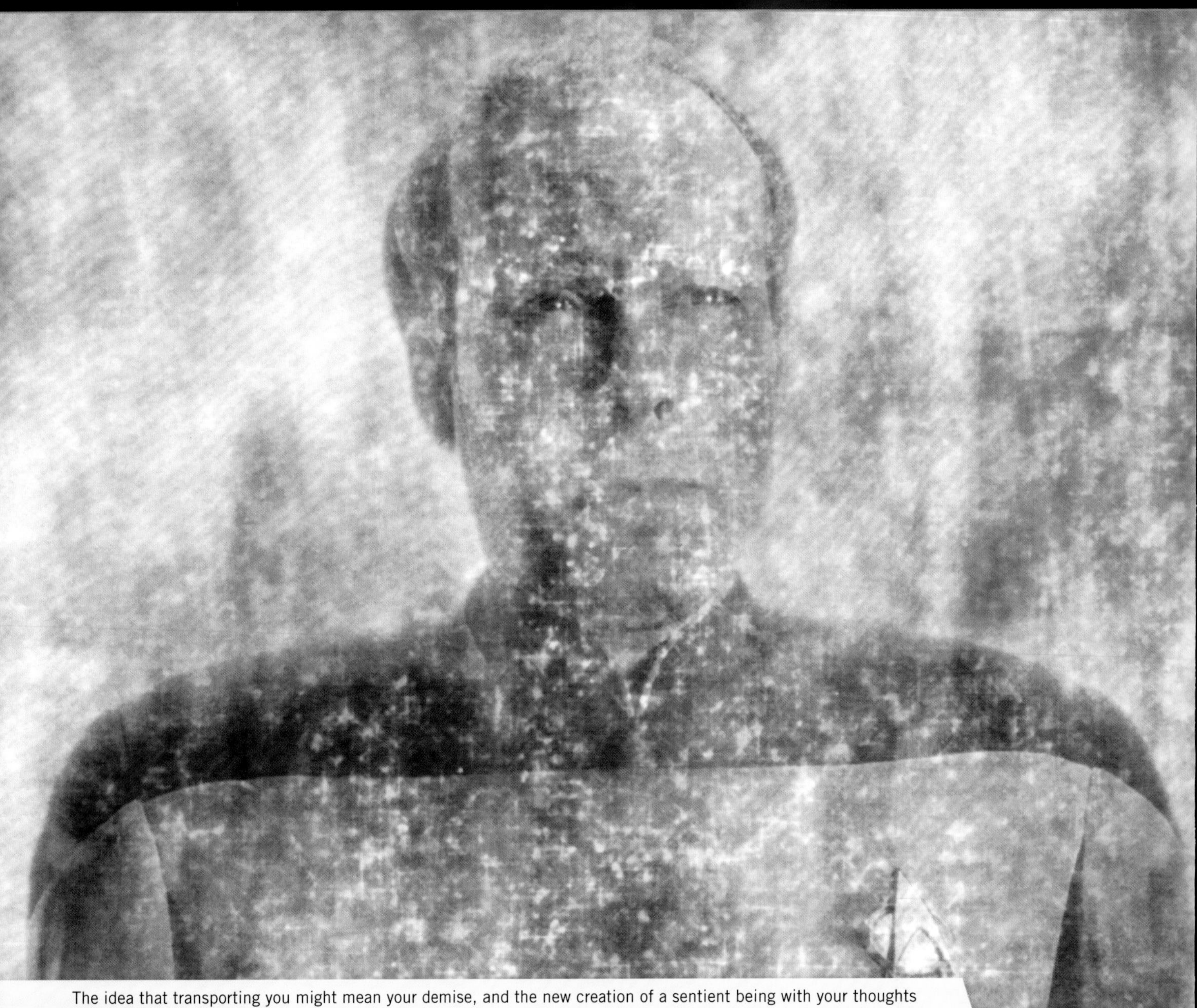

The idea that transporting you might mean your demise, and the new creation of a sentient being with your thoughts and memories but distinct from you, is one that should give us all pause before we ever step into such a device.

TRANSPORTERS

"Beam me up," orders Captain Kirk, and from the distant planet's surface, his body slowly fragments and fades away, reappearing seconds later on the ship. Not only are all his atoms intact and correctly assembled, but his thoughts, his actions, his momentum, and his exact state of being at that very instant are all perfectly maintained. While the idea of being deconstructed atom by atom and then put back together in a distant location may seem disconcerting, the possibility of near-instantaneous teleportation from one precise location to another would instantly revolutionize a whole slew of transportation-related issues, from freight to commutes to precision installation of physical components.

> ***"Gentlemen, I suggest you beam me aboard."***

Transporter technology—and general confidence in its use—improves tremendously throughout the *Star Trek* timeline, from Captain Archer's twenty-second-century proclamation that he "wouldn't even put his dog through this thing" to routine, frequent travel by transporter in the twenty-third century of the original series to most crew members regarding it as safer than any other form of travel by the twenty-fourth century in *Star Trek: The Next Generation*. (There are some exceptions, of course, such as Dr. Pulaski's abject refusal and Ensign Barclay's downright terror at travel by transporter.) Transporter accidents, a common, if tragic, occurrence in the twenty-third century, are so rare in later eras that the

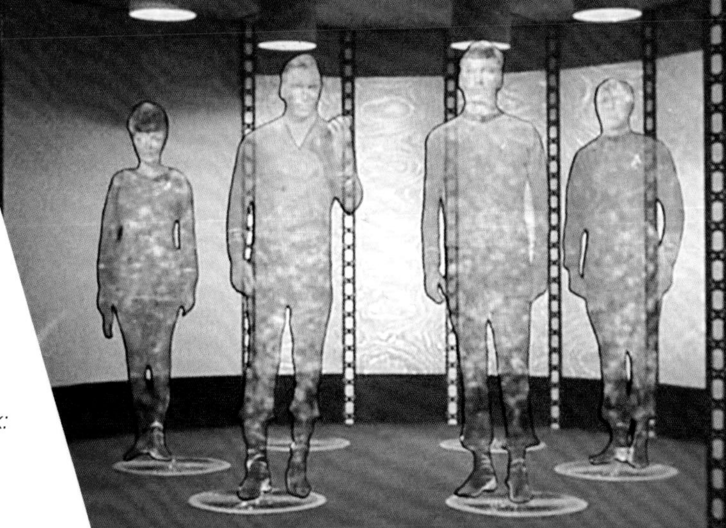

With a transporter, you could—in the span of seconds—be disassembled and reassembled at two locations up to tens of thousands of kilometers apart.

transporter accidents that created Thomas Riker in 2361 and that de-aged members of the *Enterprise*-D's crew in 2369 are the only two known imperfect transporter incidents in the entire decade.

However, if you find yourself cringing at the prospect of being deconstructed particle by subatomic particle, you're not alone. It gets worse when you consider that the atoms that made you up at your origin may be distinct from the atoms making you up at your destination in a transporter. The idea of the science behind transporters is to take the approximately 10^{28} particles composing a living human being, reading the quantum state of every particle and its interaction with every other particle in the system, and then reconstructing that exact same state (with the same number and types of particles) at the destination. The "destination" version of you will never know the difference, because quantum particles themselves are not special in any particular way—any two electrons in the

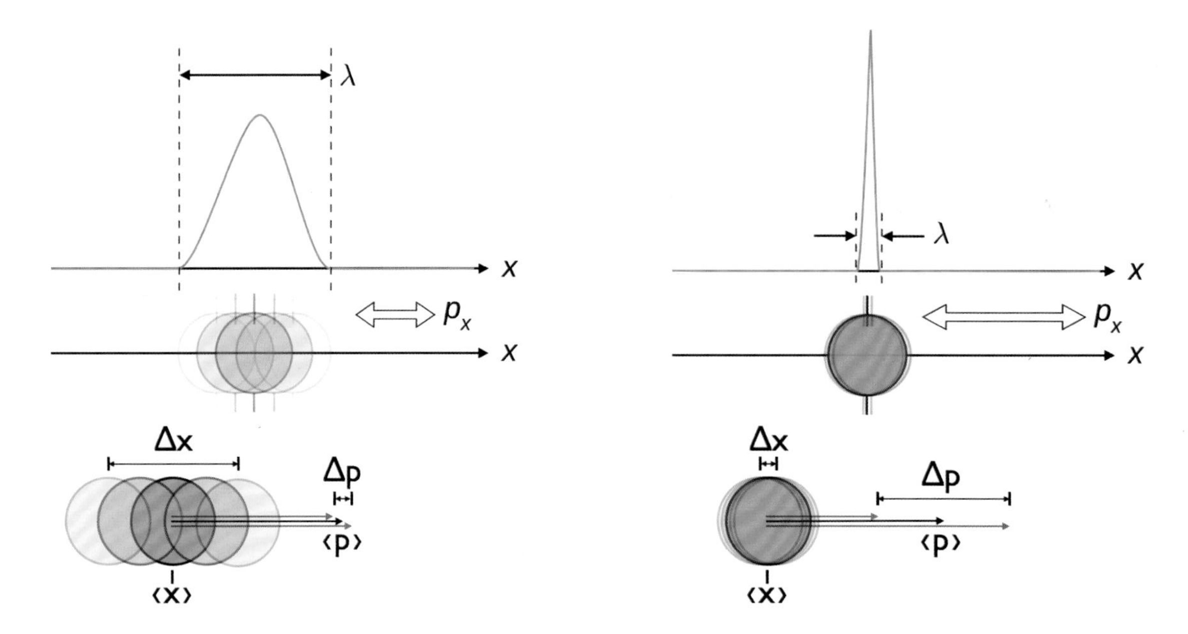

The image here illustrates that neither position nor momentum can ever be measured exactly, but that both are inherently uncertain, with a small uncertainty in one necessitating a large uncertainty in the other.

The de-aged condition of Picard, Guinan, and others was only temporary, as they were successfully re-aged through the transporter as a remedy.

same quantum state, for example, are completely identical to one another—but it is possible that the "origin" version of you would cease to exist, with only the duplicate version of you continuing onward in space and time.

To even dream of doing that would require being able not only to put all the particles that make you up back together in the same configuration, but to replicate the same positions and momenta that they had before you were teleported. Think about the difference between a living human and a corpse of a human: there are no particles that are necessarily different; it's simply the way those particles are positioned, interacting, and moving in that configuration. There is an inherent uncertainty between position and momentum as well—the Heisenberg uncertainty principle—that forbids you from knowing the complete set of information about even a single particle simultaneously; there's an inherent indeterminism in nature itself! But this isn't, from a pure physics point of view, necessarily the end of the story.

That is because what you can do is transfer an arbitrary amount of information from one location to another through the process of quantum teleportation. The name is a bit of a misnomer, since this isn't the teleportation of actual quantum particles, but of the information about the states of quantum particles. Make enough pairs of entangled particles between two different locations, and you can teleport that information from one location to another: you can move the state and the information of one object from point A to point B without having to move the object itself. This discovery was made in 1993 by a team of physicists who proved that the entire quantum state of a system, including entanglements, can be transferred without exchanging any matter between the two locations. It's possible that combining this technique with the emerging technology of quantum computing could enable the total information necessary to encode a living human being to be scanned in and teleported from one location to another. Since the source of the information prior to transport doesn't need to be destroyed, you could even use it to clone a person in their exact state at a specific, critical moment!

The main challenge of a transporter, however, is reconstructing that matter on the other end. Knowing the information state of a human being—including all their component particles—is one thing, but reconstructing that human being is quite another thing entirely. Despite Russia launching a $14 trillion program, the National Technology Initiative, that includes the goal of teleporting a human being by 2035, it isn't clear that this part of the technology is feasible given our current understanding of physics. And while quantum tunneling is a real phenomenon, enabling a particle to reach the other side of a barrier it doesn't have enough energy to hurdle or break through, the probabilities are so exponentially suppressed that macroscopic objects cannot move an appreciable distance through space via this process. A single atom of modest energy might spontaneously tunnel through a thin sheet of paper a few times out of a million attempts, but a tennis ball never will.

There are many different ways one can conceive of a teleporter. As it turns out, *Star Trek*'s vision, of transporting the information and reconstructing the matter in that desired configuration at the destination, may actually be physically feasible. In fact, of all the puzzles that this technology requires to be solved, the transportation of the full suite of information encoded in a human being may be the first one to fall. Actually determining the quantum state of the source of a human being desiring transport and reconstructing the matter at the end may be even more difficult. If you replace one of your original atoms with another, it's clearly still you; what if you replace half of them, three-quarters, or even a hundred percent of them? Ensuring that the human being that arrives has the same "you-ness" of the one you sent through the transporter is a difficult enough problem that not even paranoid genius Reg Barclay addresses it. Although this technology falls firmly into the speculative category, there have been substantial enough advances in theoretical and experimental physics that indicate it may be possible after all. But until the "you-ness" problem has a clear-cut solution, we might want to limit our transporting to inanimate objects.

If you can move a quantum state and the information contained therein from one location to another, it should be possible to copy that state as well, resulting in a transporter duplicate—as in the case of Will and Tom Riker.

Impulse engines have always been used for precision maneuvering, from as far back as the maiden voyage of the *Enterprise* NX-01 up through all the starships of the twenty-fourth century.

IMPULSE ENGINES

Warp drive may be indispensable for traveling from one star system to the next, but it's lousy for small-scale, precision motions. For traveling within a solar system, rendezvousing with another starship, docking at a starbase, orbiting a planet, or completing intricate tactical maneuvers, control is far more important than raw speed. That's where the impulse engines come in. Capable of taking starships up to a significant fraction of the speed of light—making interplanetary journeys a matter of hours, rather than months—impulse engines can also enable a ship to reach its desired location with submeter accuracy.

"Slow to impulse."

It was the birth of rocketry that led humanity into space in the first place. While it was Zefram Cochrane's warp flight that brought Earth to the attention of the Vulcans and enabled humanity to travel interstellar distances, simple rocketry was still used at all points in the *Star Trek* timeline to travel within a planetary system. By operating on the simple concept of Newton's third law—that every action has an equal and opposite reaction—a starship can be made to accelerate by imparting large amounts of energy to massive particles that are then ejected opposite to the direction a starship will accelerate. While twenty-second-century impulse engines could accelerate a ship up to approximately 5% the speed of light, larger, more powerful ships in the future, like the *Enterprise*-D and *Voyager*, could travel five to ten times as fast under impulse power.

The two components of any potential impulse engine are the generation of energy and the expulsion of high-speed, massive particles. It's important to generate large amounts of energy to kick those massive particles out with the maximum possible amount of momentum, and then the equal-and-opposite reaction is an equal-and-opposite change in momentum of your starship. If you want to accelerate a starship to any appreciable amount of the speed of light, that means imparting a huge mass with a huge change in velocity. Under the laws of physics, the particles that come out of the exhaust—which are presumably much lower in mass than the starship itself—must be moving incredibly close to the speed of light. In physics, momentum has to be conserved.

But energy, too, needs to be conserved! There must be some way to generate enough energy to make a ship move that fast, and to accelerate those exhaust particles so close to the speed of light to make that happen. These energies aren't so easy to come by, physically; accelerating just a kilogram of mass up to 5% the speed of light takes 113 terajoules of energy. That's the equivalent of 27 kilotons of TNT, or more energy than was released in the Hiroshima or Nagasaki bombs. To accelerate the *Enterprise*-E—estimated mass: 3 million metric tons—to that same speed, would require approximately the amount of energy that was released in the asteroid impact that wiped out the dinosaurs. You would have to annihilate about 4,000 metric tons of matter and antimatter, or burn through about 1 million metric tons of nuclear-powered fuel, to release that much energy. And that's assuming you can transfer the entirety of that energy into accelerating the high-energy particles in the proper direction out of the ship's exhaust.

While *Star Trek* claims to use deuterium fusion to power the impulse engines, burning through a third of the ship's mass just to go from rest to full impulse seems like a bad long-term plan for a starship. Since mastering matter-antimatter power (as part of antimatter containment; see page 43) is required for *Star Trek*, perhaps the energy needs can legitimately be met. In addition, channeling that energy into momentum isn't so difficult. In fact, if the matter-antimatter power is made of electrons and positrons—which annihilate purely into photon energy—they can be shunted out in the opposite direction from how you want your ship to accelerate, and off you go at full impulse. You're limited only by the efficiency with which you can reflect the photons and use their momentum to power the ship. If you want to go the route *Star Trek* claims and use a plasma exhaust instead, it's theoretically possible, although it's also less efficient as far as momentum transfer is concerned. Every additional collision or transfer of energy is one more inefficiency in the chain of events.

The last obstacle to overcome is a real challenge: the issue of accelerating the ship from rest to a significant fraction of the speed of light could have catastrophic consequences for the crew. Human bodies can't handle accelerating at speeds of more than g, the gravitational pull of the Earth at the surface: 9.8 meters per second squared. But if you accelerate to full impulse, even if it takes you an hour to get there (and it never takes that long), you'll need to accelerate at a minimum of 4160 meters per second squared, or 425 g. The only humans to ever survive accelerations of that magnitude have been in crashes, where that level of g-force lasted for less than 0.1 seconds, not for an hour. Some type of

A starship, more than half a kilometer in extent, can pass through an opening barely larger than itself without any risk of collision under impulse power.

The first test firing of SpaceX's Raptor engine on September 25, 2016, illustrates how the exhaust must be expelled antiparallel to the direction of the desired thrust.

"acceleration dampener" would be needed for a human to survive the acceleration created by an impulse engine, otherwise the blood would fail to circulate through the human body. Trying to overcome this with a larger acceleration in a shorter amount of time would be even worse, as the impact would rip the blood vessels right out of your vital organs, similar to what a skydiver whose parachute doesn't open experiences. Acceleration dampeners—equivalent to a gravity shield or *Star Trek*'s inertial dampers—are thought to be physically impossible, however; they would violate Einstein's equivalence principle, which is the foundation of general relativity. Any sort of gravity shield, mandatory for human survivability aboard a rapidly accelerating starship, is purely hypothetical.

There are real advances happening surrounding engine-based technologies that could potentially lead us on interplanetary voyages in greatly reduced timescales in the near future. Chemical-based rockets, which convert only ~0.001% of their mass into energy, are presently used, but may be superseded by nuclear rockets, which can be nearly one thousand times more efficient.

Deuterium-lithium fusion rockets are under development today, and may cut down the journey to Mars from nearly a year to just over a month. High-energy plasma drives (with exhaust) may not quite reach photon-levels of efficiency, but represent significant improvements over current technology. An out-of-the-box technology, a laser sail, may enable one-way acceleration to near-light speeds of very lightweight robotic probes, as they will not be limited by the acceleration thresholds of a human body. The ultimate in efficiency—matter-antimatter power—has been demonstrated to be feasible in principle, and could someday create a true *Star Trek* impulse drive. But that technology is likely centuries, rather than decades, away.

The plasma rocket, a type of electromagnetic thruster, may be the first *Star Trek*–like breakthrough to come to fruition. It is being explored by a variety of teams engaged in plasma propulsion research, including NASA's Variable Specific Impulse Magnetoplasma Rocket (VASIMR).

Nevertheless, impulse engines are technologically feasible based on what we know of physics today. The big challenge will be the dual trade-offs: between fuel and weight, and accelerations and human tolerance. If we can develop more efficient fuel-based technologies, such as nuclear or matter-antimatter sources, our interplanetary journeys will become much faster and human-accessible, even well before they reach true *Star Trek* levels. The acceleration limits are more constraining, but so long as there's either only inanimate cargo on board—which might include humans either frozen or suspended in a fluid environment—a rapid acceleration might not be the deal breaker Einstein's theory claims it to be. Regardless, impulse drive is on its way, and as our rocketry technology continues to improve, we'll find it progressively becomes more and more *Star Trek*-like to journey from world to world.

TRANSPARENT ALUMINUM

When all that separates you from the deadly vacuum of interstellar space is the hull of your starship, you'd better make sure that material is strong, durable, and impervious to leaks, cracks, and punctures. You want it to be hard but not brittle, and resistant to everything from high-speed strikes from micrometeoroids to asteroids to any weapons your enemies might fire at you. But if you're carrying people for long periods of time, particularly if yours is a peacetime ship carrying thousands of civilians, perhaps being inside an opaque metal box isn't the most ideal way to travel through our breathtaking galaxy. Perhaps, instead, you'd want a series of viewports transparent to visible light. We have plenty of transparent materials that we commonly use on Earth for a variety of purposes—plastics, acrylic, and silicate and borosilicate glasses, for example—but they don't meet the hardiness requirements that a spaceship would demand. For that, you'd need a special type of material created specifically for use in outer space, and *Star Trek* envisioned exactly that with the idea of transparent aluminum.

Theoretically developed in the 2130s, transparent aluminum was a far superior successor to plexiglass, many times as hard and dense, enabling it to be far thinner for the same protective applications. Viewports abound in all the *Star Trek* series and films: just as Kirk shared a view with Odona, Data shared one with the Romulan admiral Jarok and Sisko shared one with Kai Opaka. Captain Archer even had a viewport in his ready room. Most famously, Commander Scott used the formula for transparent aluminum as a bartering chip in order to obtain large amounts of plexiglass—its predecessor—in *Star Trek IV: The Voyage Home*, to house the whales they needed to bring back to their own time. (In addition to being the best known reference to transparent aluminum, *Star Trek IV* was also the first time transparent aluminum appeared in the *Star Trek* franchise.) This technology was widespread enough in its use that viewports of the *Enterprise*-D, and presumably all *Galaxy*-class starships, were made of transparent aluminum.

As fanciful as this sounds, it was well known long before any of the *Star Trek* franchises were launched that aluminum could be alloyed into a transparent state. The rock-forming mineral

Transparent aluminum, or some other type of light, strong metal, would be necessary for large windows with views out into the universe on a starship to be possible (top). An alternative option is a video screen (bottom), which comes with its own challenges.

corundum, made of aluminum and oxygen atoms, has been known since ancient times, most commonly in its forms with impurities that turn them either red or blue: rubies and sapphires. The crystalline structure that aluminum and oxygen make when bound together into corundum has a series of incredible properties:

Corundum comes in a variety of colors, tints, and transparencies, but could be completely transparent without impurities.

- It's naturally transparent, having colors only when impurities are present
- It's incredibly hard, surpassed only by moissanite (silicon carbide) and diamond (crystallized carbon) among naturally occurring minerals
- It's extremely dense, at 4.02 grams per cubic centimeter, which is highly unusual for a mineral composed of such light elements in the periodic table

While corundum crystals measuring more than 2 cubic feet in size have been found in nature, the molecular structure and configuration is more useful as a design guide than as a source of raw materials for use in manufacturing.

By controlling the crystallization process at the molecular level, starting from a seed crystal that grows either spontaneously or by an induced mechanism, a general class of materials known as transparent ceramics can be produced. For a long time, both glassy and crystalline ceramic materials were only transparent in thin coatings, particularly if they were aluminum-based. Because they were constructed of fine powders that would yield a fine-grained crystalline structure when fused together, there would be scattering centers throughout if the material was much thicker than the wavelength of visible light. But recent nanotechnology breakthroughs have made it possible to create much thicker transparent ceramics, including the aluminum-based transparent spinel ($MgAl_2O_4$), alumina (Al_2O_3), and yttria alumina garnet ($Y_3Al_5O_{12}$). This class of materials originally was applied for use in lasers in the 1960s, due to the compounds' transparent nature and solid-state properties, but in the 1970s, the idea of transparent armor started to be tossed around as scientifically feasible.

In order to protect sensitive devices—ranging from planes to submarines to tanks and the fragile humans inside of them—against ballistics, fragmentation, and punctures, optically transparent materials that do not distort at the molecular level when impacted are needed. Normal transparent armor consists of layers of cheaper, conventional glass interspersed with thin layers of ceramic. But if you want progressively thinner, lighter, and higher-performing materials, you want a material that's either fully ceramic or made from many thin layers of either glass or plastic with the ceramic making up a substantial fraction of the overall material. In 1980, six years before *Star Trek IV* premiered, the first transparent aluminum ceramic—aluminum oxynitride—was produced. Although it was far too expensive for commercial applications, it had many intriguing properties:

- It was 85% as hard as sapphire
- It was more than 80% transparent to visible, near-ultraviolet, and near-infrared light
- It remained stable up to temperatures of 2,400 kelvin
- It was consistently able to resist the impact of multiple successive .50-caliber bullets

Conventional bulletproof glass was a tremendous advance, but is not as hard, durable, or transparent when put to the test as many applications, including space travel, would require.

This generic class of materials, known generically as AlON due to its chemical formula—$(AlN)_x \times (Al_2O_3)_{1-x}$, where x is between 0.30 and 0.37—is both optically and materially superior to all plastic- and glass-based alternatives. Only variations on spinel and sapphire, both of which are also aluminum-based, are competitive. Polycarbonates, acrylics, and plastics are still used for a wide variety of industrial and military applications, including bulletproof glass, due to their low cost, but AlON is far more effective and efficient than these other materials, with 2.5 times the stopping power of conventional bulletproof glass. Because the transparency of materials like AlON, sapphire, and spinel extends into the infrared, they're

The pilot's dome of the F/A-18 Hornet, shown here, is made out of polycarbonate material due to cost. A portion of the Sidewinder missile being loaded onto it, however, is made of a form of transparent aluminum to enable its heat-seeking capabilities.

also useful for applications like heat-seeking missiles, including the U.S. military's Sidewinder missiles.

Other tricks have been used, more recently, to create a temporary state of purely transparent metals, including aluminum. By using a pulsed laser to remove electrons judiciously and temporarily, aluminum can be made transparent to visible light. However, this state lasts less than 1 picosecond (a trillionth of a second) before returning to its original, opaque state, and is not as hard or dense as the composite materials listed above. If the goal is to fabricate transparent materials for use in armor, construction, or protection from a dangerous enemy or environment, a dense, hard, impact-resistant material with the right optical properties is absolutely necessary. It's remarkable that *Star Trek* correctly foresaw aluminum being the key element at play in all subsequent incarnations; an aluminum-free transparent material—such as silicon or boron, for example—has yet to supersede aluminum based on these criteria. Other elements conjectured to be critical in laboratory-based transparent metal experiments, silver and hydrogen, don't appear to work in this context; aluminum is indeed the key element.

There is a unique formula for transparent aluminum, which *Star Trek IV: The Voyage Home* revealed, if ever so briefly.

As our ability to manipulate matter at the molecular, nanoscale level improves, and as we discover novel configurations of the same handfuls of elements, it's no surprise that we've made tremendous improvements to the hardness, transparency, and durability of materials over the past fifty years. But the biggest advance has come in manufacturing and costs to produce this type of material; it would be shocking if AlON (or closely related materials) weren't in widespread use by the end of the century in everything from armored cars to fighter jets to submarines to spacecraft. It may never become as cheap as plexiglass to manufacture, and the difference between today's transparent aluminum and plexiglass may only be two to three times better in terms of thickness, rather than the four to five predicted by *Star Trek*. But it's still pretty incredible how far we've come toward a technology that wasn't supposed to come to fruition until the 2130s!

ANTIMATTER CONTAINMENT

"We are losing antimatter containment." Those five simple words, uttered by Lieutenant Commander Data moments before the *Enterprise*-D's destruction, are harbingers of nature's most destructive power: matter-antimatter annihilation.

Yet the reason antimatter containment is so important is that it makes possible the ultimate energy source, a 100% efficient conversion between mass and energy as governed by Einstein's $E = mc^2$. A matter-antimatter reaction is necessary to power a ship's warp core, as the massive but controlled release of energy it offers is unmatched by any other source, even in theory. Due to this property of

The most important safety concern for antimatter is to keep it completely separated from any normal matter—down to the smallest subatomic particle—it might interact with, as *Star Trek*'s antimatter storage pods were designed to do.

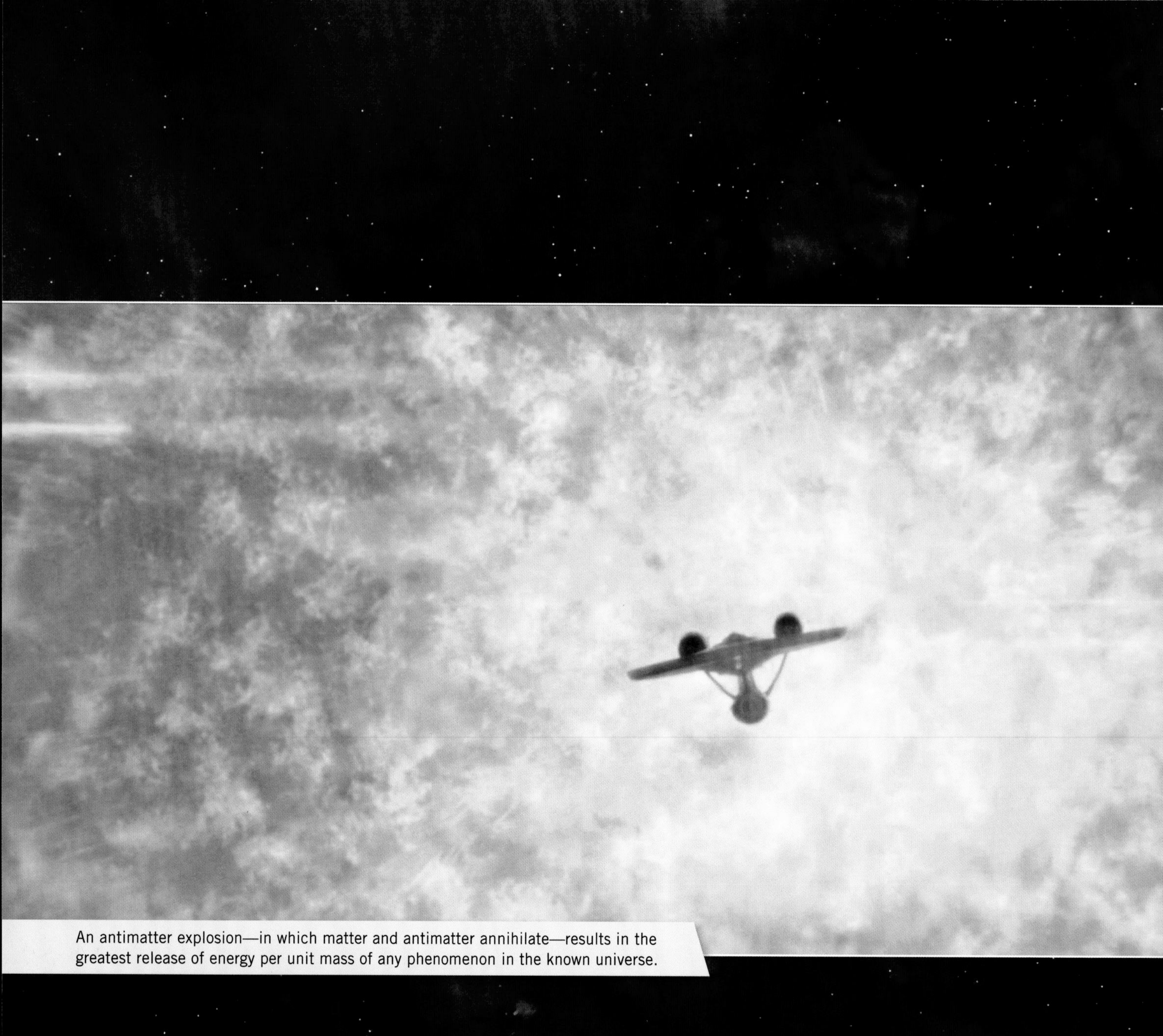

An antimatter explosion—in which matter and antimatter annihilate—results in the greatest release of energy per unit mass of any phenomenon in the known universe.

antimatter, it's absolutely imperative to keep it completely segregated from anything it might react with, which means it must not come into contact with any form of normal matter when contact isn't designed to release energy.

If the ability to produce and store antimatter in arbitrary amounts were achieved, the technological advances would be astounding. Not only would our energy needs as a planet be easily met, but the energy would be completely clean, producing no radioactive or environmentally detrimental by-products. Antimatter itself could be easily weaponized simply by colliding it with normal matter. The interaction of just a few grams of antimatter with normal matter would cause as much destruction as any atomic bomb ever constructed by humanity. Any reaction requiring huge releases of energy all at once would instantly have their needs met. The leaps forward in our society would be immediate and tremendous. Yet the volatility of antimatter ensures that in order to achieve this technology, a reliable method of containment is absolutely essential.

Producing antimatter is easy enough from a physics point of view—simply collide any two particles together with sufficient energy, and you'll have the possibility of spontaneously creating new particle-antiparticle pairs in addition to the raw materials you started with. This mechanism can produce any combination imaginable so long as the kinetic energies are high enough to allow it: electron-positron pairs, proton-antiproton pairs, or any other particle-antiparticle combinations allowed by the standard model of particle physics.

But merely producing antimatter particles represents only a tiny fraction of the battle. Without further intervention, this antimatter will collide in short order with another matter particle, as a result being annihilated the same way it was produced: through the power of Einstein's $E = mc^2$. Individual antimatter particles can be temporarily stored as they were at Fermilab, the U.S. Department of Energy particle physics laboratory: by accelerating them to large velocities and keeping them moving in a ring with electromagnets. However, the vast majority of antimatter produced and stored in this fashion at Fermilab eventually collided with the beam pipe it was stored in, making this a dangerous and ineffective method for long-term storage of large amounts of antimatter. If you want to store antimatter long-term or in large amounts, you need to create a nonreactive, electrically neutral version of it. You need to create neutral antiatoms, such as antihydrogen. Instead of protons (or

Firing two protons into one another at high enough energies means that sometimes, an additional proton-antiproton pair will be produced. From the 1970s through the 2000s, this was the mechanism scientists at Fermilab relied on to produce antiprotons for what was then the world's most powerful particle accelerator: the Tevatron.

protons and neutrons, for heavier nuclei) and electrons, you'd have to build your antiatoms out of antiprotons (or antiprotons and antineutrons) and positrons, to create neutral antimatter.

Although the existence of antimatter and the idea of its properties and applications had been theoretically known since the work of English physicist Paul Dirac in the 1930s, bound antiatoms were only created in a laboratory setting for the first time in 1995. This was achieved by bombarding heavy elements with high-energy antiproton beams, which would produce electron-positron pairs; a tiny fraction of antiprotons would bind with the newly created positrons, creating neutral antimatter. In order to attempt to contain it, however, antiatoms would have to be created *cold*, or at nonrelativistic speeds. The first long-term trap of antihydrogen was successfully carried out in 2010 by the Antihydrogen Laser Physics Apparatus (ALPHA) experiment at CERN, home to the Large Hadron Collider, in which a combination of nonuniform electric and uniform magnetic fields known as Penning traps bring antiprotons and positrons together to create antihydrogen, and then nonuniform magnetic fields are applied to bring the antihydrogen toward a magnetic minimum, thanks to the intrinsic magnetic properties of antihydrogen. By 2011, antiatoms had been trapped for nearly 20 minutes at a time.

Heavier antimatter nuclei have also been produced, including antideuterium and antihelium, but they have never been trapped or created cold the way antihydrogen has. This doesn't necessarily mean there's a tremendous technological hurdle to overcome there, however; producing large amounts of antihydrogen at the right pressures and temperatures might be enough to eventually power a starship all on its own. While we normally perceive hydrogen as a gas, at high enough pressures it begins to take on metallic properties. Gas giants such as Jupiter are expected to have liquid metallic hydrogen layers beneath the gaseous outer layers naturally, while at similar pressures but lower temperatures, solid metallic hydrogen should emerge. By certain fundamental symmetries in physics, many of which are presently being confirmed by experiments, antihydrogen is anticipated to have the same properties as hydrogen, its matter counterpart, does. If antihydrogen can be created, trapped, and brought to high enough pressures and low enough temperatures, it just might be ripe for long-term storage.

This technology is not without its dangers! Having large amounts of antimatter in close proximity to large amounts of matter always holds the potential for unwanted destruction, and the loss of antimatter

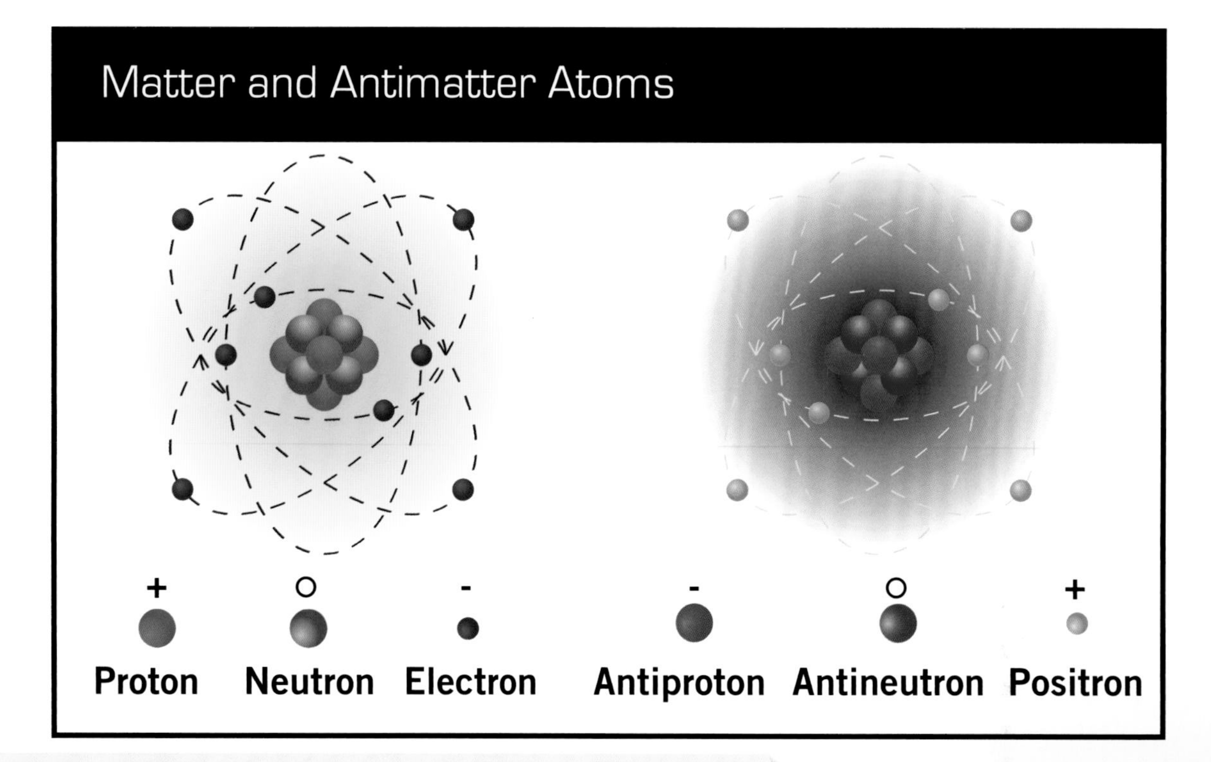

For every atom that exists in our world, it should be possible to create an antiatom counterpart. In the antioxygen atom illustrated here, protons, neutrons, and electrons are replaced by antiprotons, antineutrons, and positrons, respectively.

containment has led to all sorts of disasters in the *Star Trek* universe, including warp core breaches, the destruction of starships, and, in all cases, unstoppable, runaway annihilation of matter with antimatter. If splitting or fusing the atom was the world-changing power source of the twentieth century, where less than 1% of the rest mass of the reactants was released as energy, there's no doubt that the antiatom holds the key to the ultimate revolution in power. The prize is quite apparent: unlimited, portable sources of energy that could be released over time or all at once, with applications ranging from interstellar travel to the ultimate destructive weapon. Still, with that amount of power being stored on board a ship (with a crew!) made entirely out of matter, containment is an essential part of the equation.

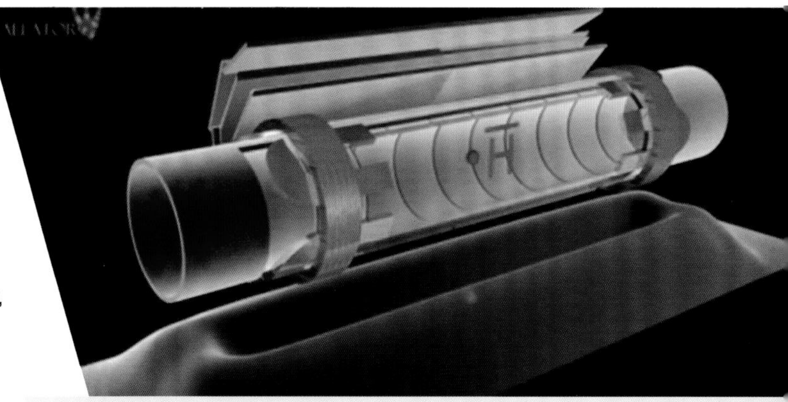

The creation and confinement of neutral antimatter in the form of antihydrogen atoms, including multiple antiatoms at once, has been achieved by the ALPHA experiment at CERN.

When *Star Trek* first came out, antimatter was physically known to exist, although only as the result of high-energy collisions. We also knew how it was created and—from the physics of $E = mc^2$—how it could be applied to create more energy per unit mass than any other material or reaction. Since the mid-1990s, however, technology has advanced to the point where neutral antiatoms can be created and stored long term, and they are presently being studied for their gravitational, electromagnetic, and nuclear properties. Unlike only a few decades ago, there's a clear path forward that's completely within the known laws of physics for bringing this technology to life. While the potential for destruction and disasters is great, this could be the ultimate solution to humanity's energy collection and storage needs, and it could usher in a truly energy-sustainable future. Plus, anytime you need a large release of energy to propel you out of a sticky situation, the decision to strategically deconfine your antimatter might be exactly what you need!

WEAPONS

AND DEFENSE

he United Federation of Planets was founded on the principles of equal rights, liberty, d freedom for all, where civilizations on different member worlds agreed to share sources, knowledge, and technology for the benefit of all intelligent, living creatures. aceful cooperation and exploration of the universe were among the highest goals of all olved, rather than exploitation, conquest, or other self-serving ends pursued by many other litical entities in the *Star Trek* universe. Yet given the existence of the warlike Klingons, the acherous Romulans, the greedy Ferengi, and the conquest-minded Cardassians, among hers, advanced military technology remained an absolute necessity for survival.

Operating under the guidelines and auspices of the Federation, Starfleet's principal functions re to gain knowledge about other planets, worlds, and civilizations; to enhance the Federation's owledge of science and technology; and to ensure the defense of the Federation and all the worlds at housed its members and allies. Just as "defense" is often a catch-all for all sorts of militaristic deavors, Starfleet starships are also equipped with a large array of weapons in addition to their fensive capabilities. Meanwhile, other empires and civilizations often employ technologies that are emed unethical by the Federation, which occasionally leaves Starfleet vessels at a tactical disadvantage en facing ships with greater weaponry, stealth technology, or the willingness to use techniques they are bidden from employing themselves.

Leaving the ethical questions aside, the Federation was never intended to police the galaxy, but merely to ive within it as its own entity. Whether facing a more, less, or equally advanced civilization, the Federation ays operated under and abided by the same rules of engagement. Yet, the Federation did not always emerge torious, and the technologies employed (or not employed) during combat often determined the outcome. spite being imagined over the past half century alone, many of these technologies—envisioned for hundreds of ars in the future—are already on their way to becoming real in the early twenty-first century.

y to render a large, macroscopic object completely invisible may have
f as science fiction, but there's actual science bringing this into reality today.

CLOAKING DEVICES

"Klingon vessel decloaking directly ahead, sir."

Even when the Klingons were allies, those words were always coupled with a heart-in-your-throat moment. More ominous were the times when Romulan was substituted for Klingon, as was often the case—including the first time a cloaking device ever appeared in *Star Trek*. The idea that an object as large and dense as a spaceship could be rendered completely invisible not just to light but to any sort of sensors, electromagnetic or otherwise, went far beyond any technology known at the time. But while it may have seemed an imaginative, unrealistic fantasy fifty years ago, this is one technology toward which science is making incredible strides.

The ramifications of being able to fully cloak an object, virtually at will, are tremendous. Just as a submerged submarine can be effectively undetectable, the reasoning went, so could, perhaps, a sufficiently advanced spaceship. The element of surprise would always be yours: passage into forbidden territory would have absolutely zero repercussions; you'd have a distinct advantage escaping an attacker, even when outgunned; and performing any feat of stealth, regardless of countermeasures, would be child's play. Short of running directly into another material object or the direct line of fire of a weapon, there would be no practical way to stop you from carrying out the vast majority of missions. If you were to couple that with malicious intent—as has often been the case in *Star Trek*—you'd almost certainly get the first shot in under any circumstances. Far exceeding mere camouflage, true cloaking would bring you invisibility and an undetectability that would give you seemingly unprecedented military superiority.

But were they truly unprecedented? By the time *Star Trek* premiered in 1966, Lockheed's Skunk Works had already been producing the A-12 reconnaissance plane for years. The first aircraft equipped with stealth technology, the A-12 was the precursor to many other stealth aircraft, including the much more famous Blackbird SR-71, although the program was only officially revealed in the 1990s. These planes' practical invisibility to radar—long-wavelength electromagnetic radiation—was the first practical step toward true invisibility.

The way to extend this across the entire electromagnetic spectrum was, in fact, not only a physical possibility but remarkably a quite straightforward proposition. Under normal circumstances, electromagnetic radiation is either absorbed or reflected upon contact with a surface. Both absorption and reflection are surefire ways to detect the presence of an object: absorption means that any background light and signals are obscured, while reflection means that any signal you send out will bounce back to you. An absorbing aircraft will be detectable as a dark spot in the omnipresent background glow, while a reflective aircraft will send a portion of the original beam back to the emitting source. Stealth aircraft are designed to minimize the reflection of radio waves as much as possible, making them difficult or impossible to see on a radar screen. But if there were a way to avoid both absorbing and reflecting radiation, instead diverting it around the object in question, that could cause the object to be effectively invisible to all electromagnetic signals—If you sent radar waves in the direction of a cloaked object, your signal would be directed around the object and continue as though it weren't there at all, while any background radiation would simply appear to originate from its normal starting point. Unless you were right on top of the cloaked object and could see the imperfections of what *Star Trek* would call a distortion field, the object would truly appear invisible.

There are, in fact, a number of known "metamaterials" that make electromagnetic radiation pass freely around an object, making a true invisibility cloak—distinct from passive and active camouflage—a real physical possibility. Unlike standard materials, which may be transparent at specific wavelengths and simply transmit light through them, metamaterials derive their properties not from their composition but rather from the way they are structured. By guiding electromagnetic radiation around them, the final signal, a long distance away, appears identical to the initial signal; to an observer, the device is truly cloaked and hidden. These metamaterials, and their associated optical properties, are not limited by the barriers of individual materials, such as glass. Under a metamaterial cloak, the object itself has no properties about it that are different, as matter running into its location still collides with it, but the incident electromagnetic waves are diverted without interacting with the metamaterial object itself. Simply applying a metamaterial coating, under the right conditions, is enough to completely cloak or hide a device of any shape or size.

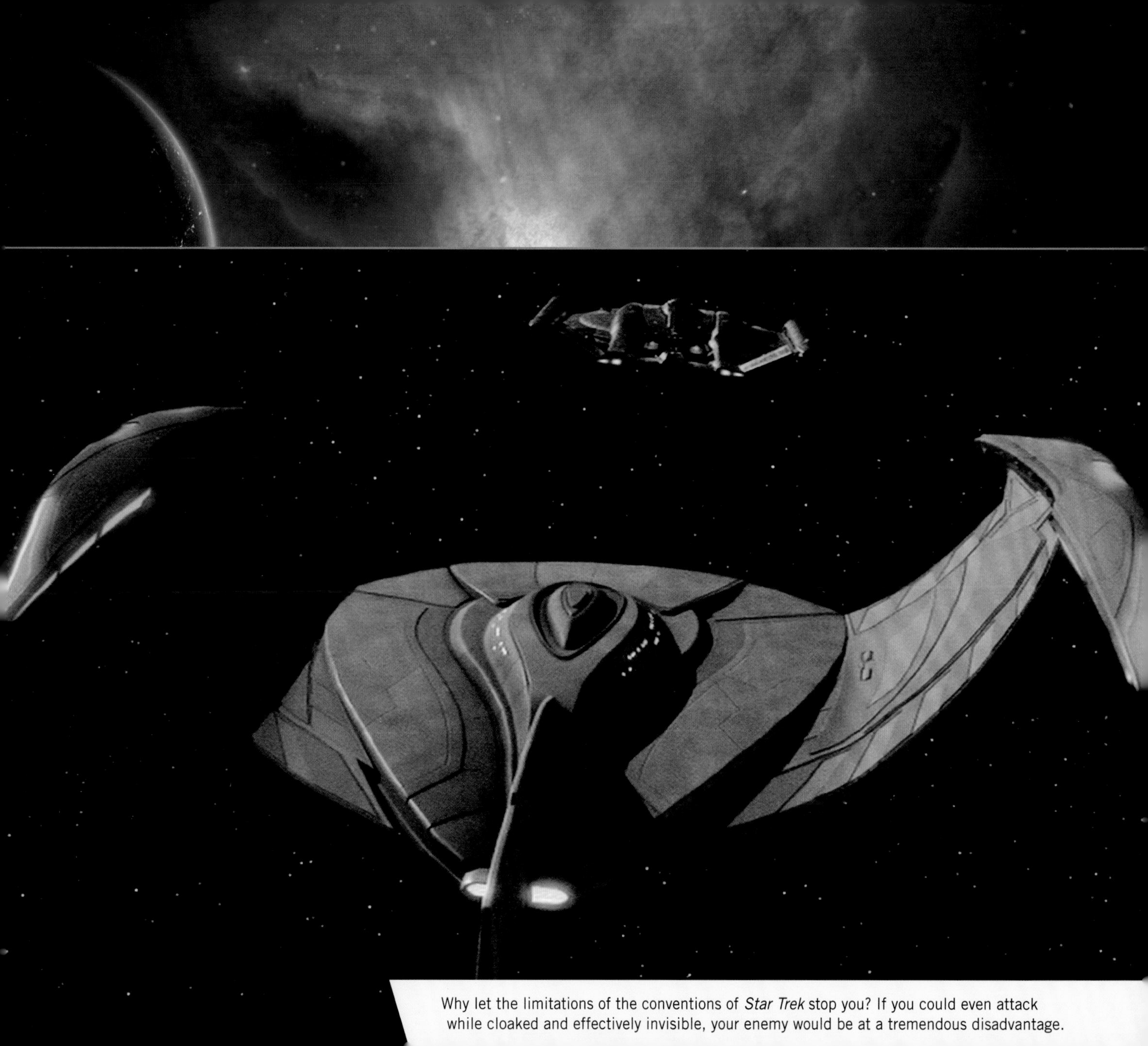

Why let the limitations of the conventions of *Star Trek* stop you? If you could even attack while cloaked and effectively invisible, your enemy would be at a tremendous disadvantage.

U.S. AIR FORCE

Beginning in 2006, the science of transformation optics allowed us to map an electromagnetic field onto a twistable, spacelike grid. When the grid gets distorted, the field does as well; if the field is twisted or distorted into the right configuration, an interior object can be completely hidden. The cloak's material must vary from point to point to render a real, physical object invisible, as the electromagnetic field must vary in order to bend (and then unbend) light by the proper amount. First successfully demonstrated in a narrow microwave portion of the spectrum, it's been proven that metamaterials can have their range of effectiveness arbitrarily extended, at least in theory, with no physical limits to the wavelengths against which they can cloak objects. Arbitrary refractive indices—stackable, customizable, and changeable at will—enable the power of these objects. The year 2016 saw the development of the first seven-layer metamaterial cloak that successfully made a macroscopic object invisible across a broad spectrum of wavelengths. However, there is a limit to the technological achievements we've reached so far: although cloaks in the infrared, microwave, and radio portions of the electromagnetic spectrum have been successfully achieved, there has not yet been a successful visible light cloak developed and employed.

It stands to reason that as metamaterial technology and its applications continue to develop, there's an excellent opportunity to extend the light-bending technology of a material's coating across the entirety of the electromagnetic spectrum. Already,

Though the A-12 reconnaissance plane didn't have cloaking capabilities, its design made it practically invisible to radar.

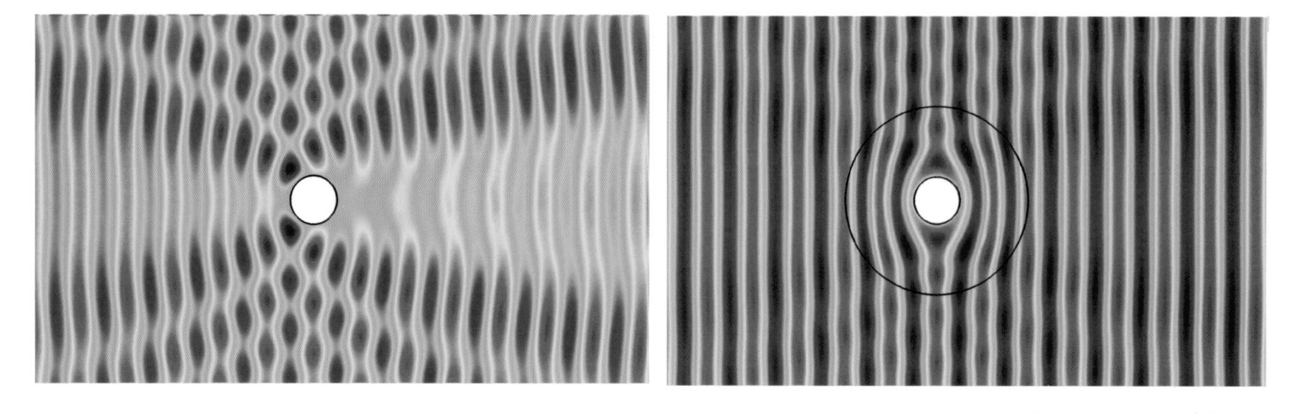

A cloaking device that was deactivated (left) would result in an object absorbing and reflecting electromagnetic signals, just as any opaque object does. But an active cloaking device (right) results in an object appearing invisible to electromagnetic signals, so long as the observer is at a great enough distance and the cloak is of sufficient quality.

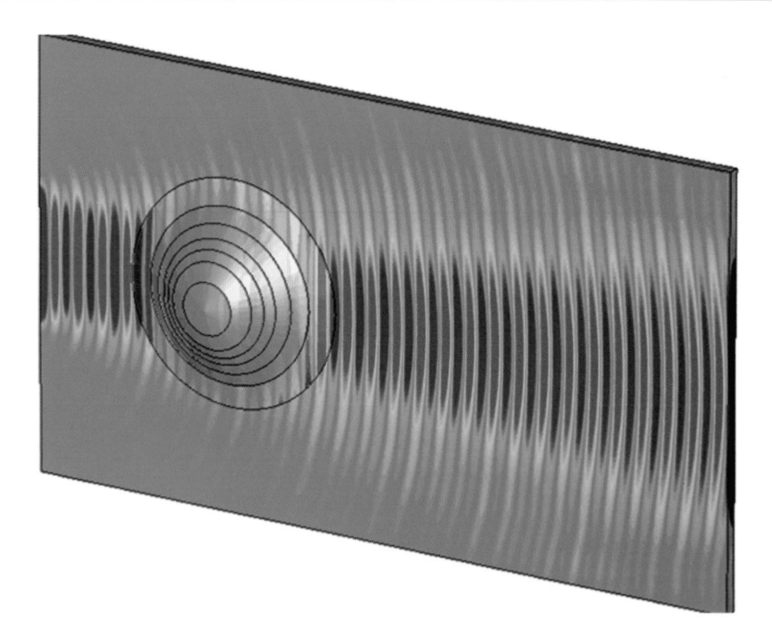

An illustration of the first successful metamaterial developed to practically cloak an object across a variety of wavelengths, with the activated cloak causing an incoming light signal to simply reappear on the other side of a solid object, virtually unperturbed.

the technology has reached a point where invisibility to long-wavelength detectors (such as radar) is becoming practical; it may not be long before human eyes can be just as easily fooled. The ability to hide in plain sight would make possible everything from enhanced natural resource exploration and underwater or subterranean archaeology to the more militaristic applications of cloaked bombs and mines. The ultimate *Star Trek* dream, of having an entire ship rendered invisible, should not be far behind. And if technology progresses on Earth as it does in the *Star Trek* universe, it likely won't be long before many competing factions have access to this technology as well.

Although we have some time to wait yet, it will be a Nobel-worthy achievement when the visible portion of the spectrum finally yields to the advances of transformation optics. It took centuries along the *Star Trek* timeline, from *Enterprise* to the original series to *The Next Generation*, for the Romulans to develop and perfect their cloaking technology, noticeable as their ship designs changed with the centuries as well. But here on Earth, it may not require centuries of development for an optical cloaking device to come to pass—we may be mere decades away.

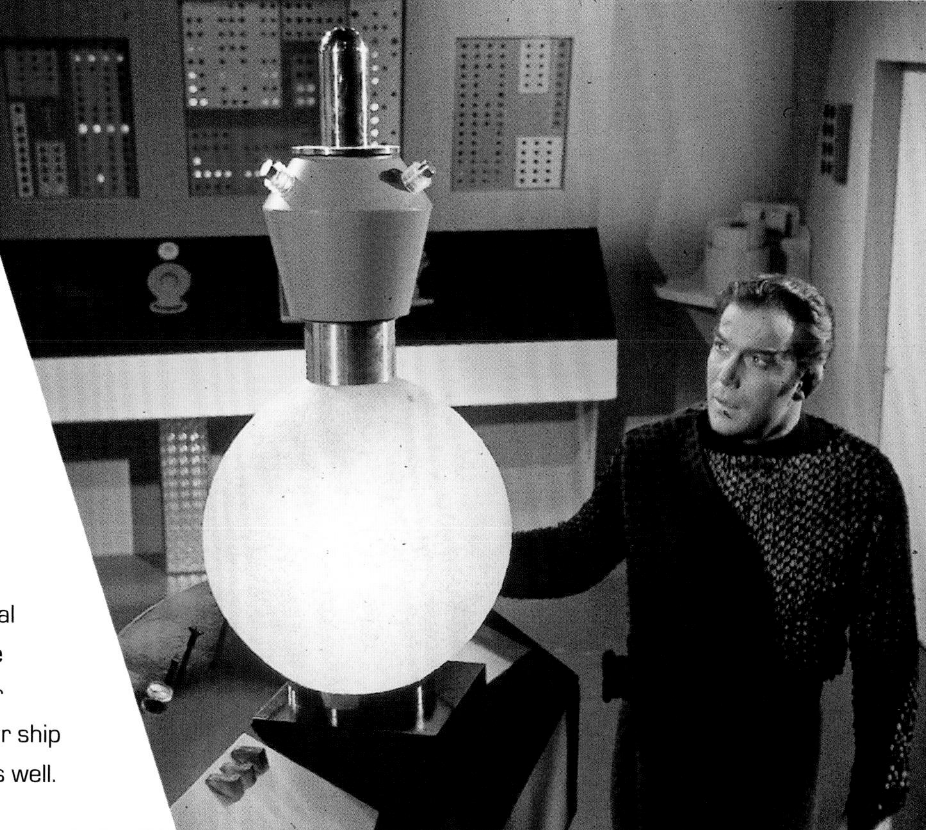

The Klingons and Romulans, among others, had full access to cloaking technology, and made liberal use of it in situations where it gave them a tactical advantage. Here, Kirk prepares to steal a Romulan cloaking device.

Deflector shields are one of a ship's first means of defense against enemy weapons.

DEFLECTOR SHIELDS

With a single command to active your shields, your starship is ready for anything from taking enemy fire to encountering a plasma storm to navigating through a debris-filled region of space. Instead of requiring a thick hull that degrades as a ship passes through the harsh interplanetary and interstellar medium, shields enable a ship to sustain absolutely no damage so long as they remain intact. In addition, they protect the crew from potentially hazardous radiation, stopping any offending particles short of encountering living tissue. Most effectively, shields are a tremendous defensive asset, allowing a ship that's fired upon to sustain minimal or no damage so long as the shields hold. Without them, catastrophic damage or even total destruction could ensue from a single direct weapon strike.

Even if a ship has sustained heavy damage, if it can take advantage of an opposing ship's shields being down to land a single well-placed photon torpedo or phaser strike, it can turn the tide of battle. In *Star Trek II: The Wrath of Khan*, the *U.S.S. Reliant* loses its nacelle in a photon torpedo strike.

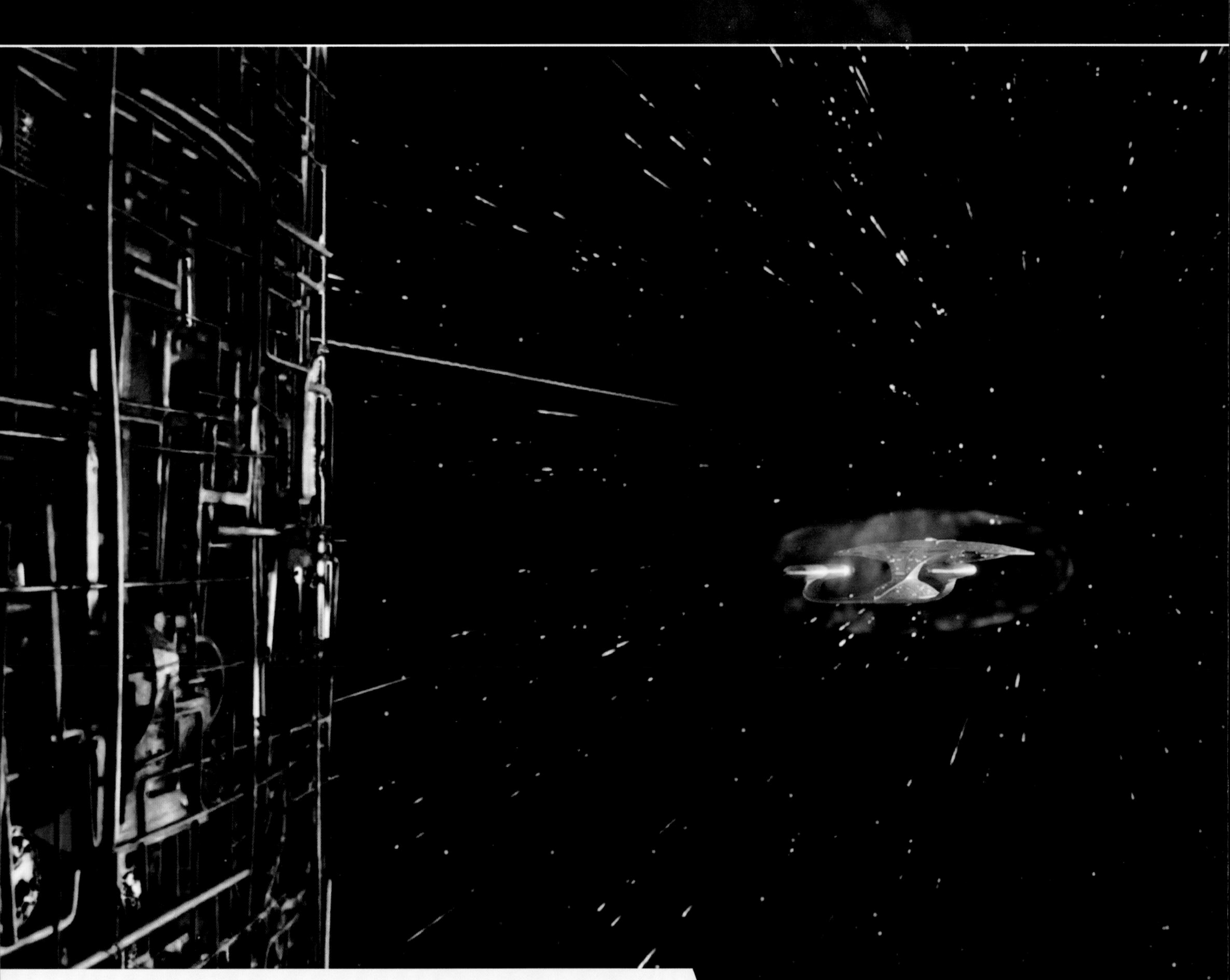

When the *Enterprise*-D first encountered the Borg, the Borg cube fired a missile that did no physical damage—but drained the *Enterprise*'s shields completely.

This difference often led to the stratagem of lowering your adversary's shields before engaging in a firefight as one of the most successful tactics in space battles.

Deflector shield technology appeared to develop rapidly in *Star Trek*'s twenty-second and twenty-third centuries, with little further improvement thereafter. The *Enterprise* NX-01 was originally outfitted with polarized hull plating, and shields were not mentioned early on, although technology provided by the Andorians eventually augmented the NX-01 with shields. By the twenty-third century, all Klingon, Romulan, and Federation starships were equipped with strong shields that could resist a direct strike from even the mighty photon torpedo. In the twenty-fourth century, however, each and every hit would still weaken a starship's shields, to the point where attacks that impacted the shields could cause damage to the ship. Eventually, if the shields failed, the ship would be defenseless.

The Hunters, a species from the Gamma Quadrant, used handheld deflector shields built into their uniforms, allowing them to receive incoming phaser fire without any notable ill effects.

Normally, neither matter nor highly-concentrated energy could penetrate through a shield, making it ideal for protecting not only starships but space stations and entire planets. A handheld version was developed by some species in the galaxy by the latter half of the twenty-fourth century.

So how do you create a deflector shield, anyway? In the world of *Star Trek*, a high concentration of gravitons is placed in an energy-containing layer (or layers) surrounding a ship, starbase, or planet, preventing matter or energy from passing through. This includes shuttlecraft, tractor beams, and transporters, meaning that shields need to be dropped to let them through. The shields require some sort of emitter, antenna, or dish, or an array of these technologies, in order to fully protect the spacecraft in question. Although normally projected around a single starship, shields could be extended around another vessel to help protect it. In addition, a graviton analysis would reveal the shape and strength of the shields around a vessel, and it's been demonstrated by the Borg that shield-neutralizing technology is possible.

Currently, scientific progress has not yet proven the existence of the graviton, much less a method of harnessing it to create an artificial shield. But if what you're seeking is protection from fast-moving ionized particles, energy weapons, and spaceship-sized objects, we already have an outstanding model of that here on Earth: the tandem of our atmosphere and our magnetosphere. The universe constantly bombards us with high-energy ionized particles from the sun (the solar wind) and from even higher-energy sources in our galaxy and beyond (cosmic rays). Earth's magnetic field, generated from a dynamo in our planet's core, doesn't just extend through Earth and cause compass needles to deflect on the surface—it extends far into space. Like all magnetic fields, it causes charged particles in motion to bend, preventing our planet from being hit by all but the fastest-moving charged particles.

The particles, photons, and large objects that do get through the magnetosphere then have to contend with Earth's atmosphere: a more than 100-kilometer-thick layer of atoms and ions. Even the most energetic particles from the sun fail to make it down to the surface, petering out dozens of kilometers up and creating nothing more than an auroral display. The high-energy photons strike matter and produce a shower of particles, similar to cosmic rays, cascading in such a fashion that only large numbers of lower-energy particles strike the surface. And macroscopic objects, like meteors, need to be incredibly large and massive to reach the surface; the vast majority of objects, up to a size comparable a large truck, will completely disintegrate in Earth's atmosphere.

But outside the protection of Earth, these charged particles become very dangerous to a potential space traveler. Someone looking to voyage even to Mars would absorb around a hundred times the radiation dose of a typical person on Earth. The simplest solution would be to create a spacecraft hull many meters thick, but that's both cost and fuel prohibitive. But in 2013 and 2014, a number of independent teams began development on the same idea: to create an ionized plasma some distance

Earth's magnetic field protects all living creatures on the surface from potentially deadly cosmic radiation by deflecting it away from our world.

away from a ship and confine it with a strong magnetic field. The field would not only deflect away incoming dangerous charged particles, but would reflect away electromagnetic radiation, similar to the way Earth's ionosphere can reflect light. Although Earth only reflects low-energy radio waves, a stronger magnetic field would correspond to a denser plasma, enabling more energetic radiation to be scattered away. With powerful enough magnets, we could literally reflect directed energy weapons, such as lasers, and protect a starship with bona fide deflector shields.

With further improvements in power and scale, it should be possible to magnetically confine an intense plasma shield around any desired object, including a spaceship. With an advanced version

To certain fortunate skywatchers, aurorae and bolides—large meteors that burn up in Earth's atmosphere—can be a beautiful sight to behold. Here, the aurora borealis can be see over Tromsø, Norway.

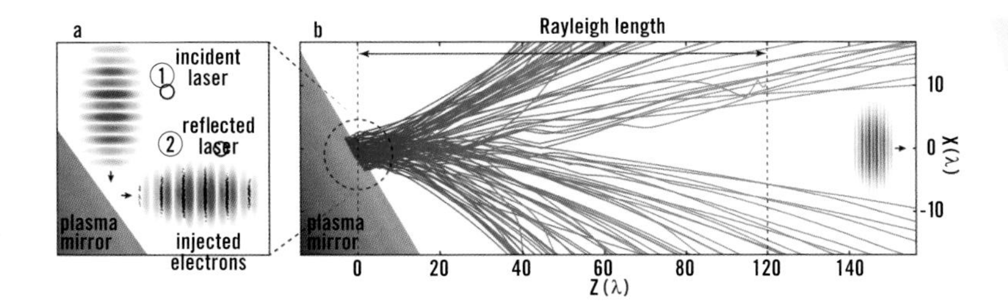

A plasma mirror, illustrated here, can reflect laser light (red and blue pulses, left) at the cost of having electrons (blue lines, right) scattered out of the plasma. This was successfully demonstrated experimentally in 2015.

of this technology in place, laser or phaser fire would be reflected rather than absorbed, and as electrons were kicked out, a weak section could develop in the shield, just as it does in *Star Trek*. Fast-moving ionized particles—normally a great danger to spacefaring, organic matter—would be deflected away, protecting the crew from one of the great dangers of space travel. And any matter wanting to pass across this plasma border would have to survive the trip through the shield itself, a difficult proposition for most matter-based spacecraft or weapons. Finally, just as *Star Trek* envisioned, many other systems would be unable to function properly with the shields up, as an intensely confined plasma would make the universe beyond the extent of the shields opaque. With a realistic deflector shield in place, not only tractor beams but technologies allowing you to see or communicate past your shields, such as scanners and even visual displays, would be knocked offline.

It's uncertain whether a single generator akin to a ship's main deflector dish could generate both the plasma and the magnetic field necessary to protect a large region of space, as it does in *Star Trek*. It appears that a more straightforward solution would involve placing multiple generators around a ship's border. This latter solution would also alleviate the potential problems associated with generating such a large magnetic field that needed to encompass an entire ship from a single source. Although it's possible that far-future developments in physics will eventually lead to graviton-based technology, the idea of a shield in space capable of deflecting energy weapons is well on its way to becoming a reality. The "plasma mirror" may only be the first step toward this dream, but it's an incredibly important one. Assuming there are civilizations out there worth meeting and exchanging knowledge with, it's only common sense to bring along some level of protection, just in case they aren't as initially benevolent as *Star Trek* might have hoped.

A well-placed phaser hit can damage or even outright destroy some of the most powerful starships in the Galaxy.

PHASERS

Ray guns and similar devices have been a staple of science fiction from the genre's earliest ventures onto page and screen. *Star Trek*'s spin on the idea—a handheld energy weapon powerful enough to disintegrate or incinerate a large, massive object yet sensitive enough to have multiple stun settings that would disable living targets but leave them otherwise unharmed—was the perfect mix of power and restraint for this fictional universe, with its focus on peace and cooperation. While handheld phasers did have kill settings, they were only used as a last resort: when maximum stun proved ineffective or in clear life-or-death situations. The energy output could be scaled up as well, in rifle-form for security teams or even outfitted as part of a starship's weapons systems.

"Okay. Make nice. Give us the ray gun."

To someone unfamiliar with twenty-third-century technology, a "ray gun" is as good a name as any for a weapon with the capabilities of *Star Trek*'s phaser.

Although the phaser itself didn't exist in *Star Trek*'s history until the twenty-third century, its predecessors, such as phase pistols and phase cannons, were similar in that they were directed energy weapons. In addition, laser-based weapons were in use by many groups, and the crew of the *Enterprise* NX-01 encountered them on a regular basis. By the time the timeline-splitting battle of the *U.S.S. Kelvin* and the

Narada occurred, ships were equipped with phaser banks, while later on (at least back in the prime timeline) the phaser system itself was tied directly to the warp core, further increasing their power capacities. And in the twenty-fourth century, the technology had evolved to the point that ships all the way up to *Galaxy*-class vessels had multisegmented phaser arrays with the additional option, beginning with the *U.S.S. Defiant*, of rapid-fire phaser cannons.

In one form or another, phasers have been the go-to reusable, rechargeable choice for hundreds of years in both handheld and ship-based weaponry, ranging from mild stun settings all the way to catastrophic disintegration or vaporization outputs. The idea of a directed-energy weapon is nothing

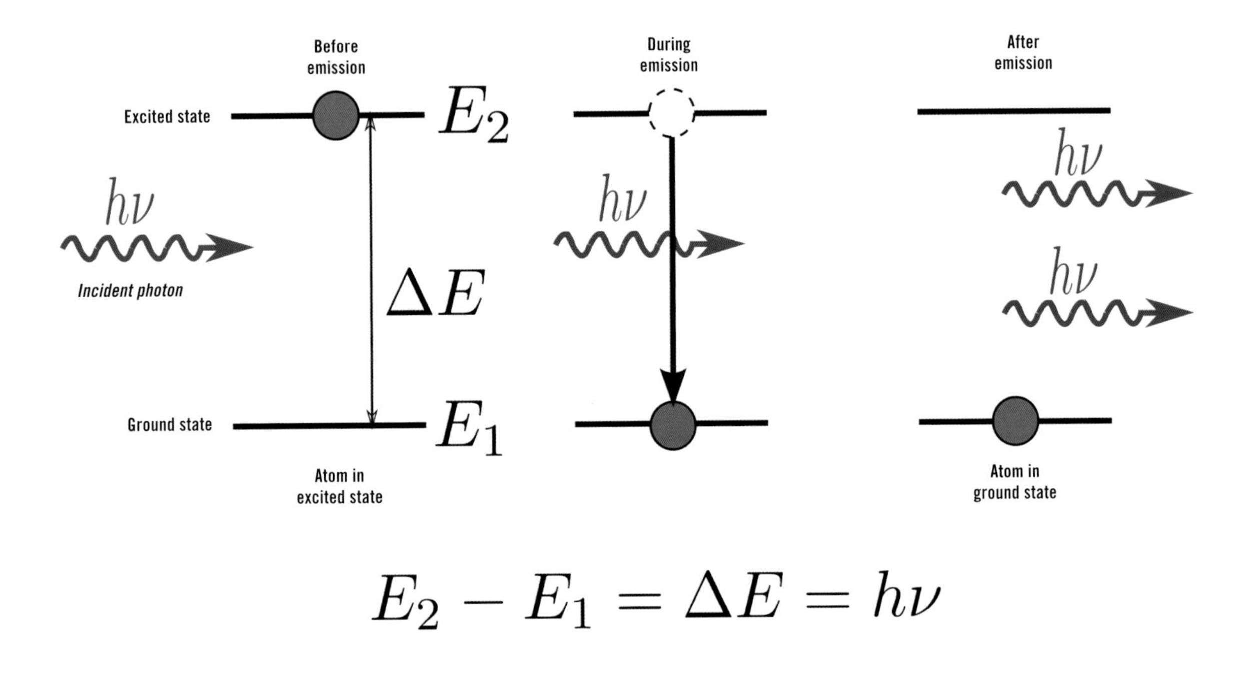

Coherent light, such as that from a laser, has a myriad of applications, including for weaponry, that normal, incoherent light is simply no good for.

new, as there were many prototype World War II weapons that had been developed, ranging from sonic cannons that could disable targets by vibrating their internal organs to electromagnetic weapons like microwaves and X-rays that imparted large amounts of energy to whatever they hit. More modern devices include variations of a weapon called a dazzler, which overloads a target's visual cortex and causes temporary blindness. At the time *Star Trek* premiered, energy weapons capable of disabling someone, knocking them down (and possibly knocking them unconscious), but leaving them otherwise unharmed didn't exist. In fact, in 1966, even the laser was a little-known device, having only been invented six years prior.

The physics of the laser was the impetus for the *Star Trek* phaser, although it was the better-known microwave version from astronomy—the maser—which was likely the true inspiration. Normally, light comes in a variety of frequencies, energies, polarizations, and directions, like the sunlight we're familiar with. But specific configurations of atoms or molecules will only emit light of a very particular frequency, corresponding to an atomic or molecular transition. If you can excite an atom or molecule using light, heat, or kinetic energy and then have it emit at that frequency in a very particular fashion, you wind up with laser light: in-phase, coherent, and synchronous.

Using the physics of the laser as a jumping-off point, humanity has now created the technology for a pulsed-energy projectile in the real world. First developed by the U.S. military in the early twenty-first century, directed-energy weapons send a pulse of nonvisible light toward a target; upon impact, it creates a small amount of exploding plasma. By sending a two-phased pulse, the early phase causes electrons on or near the target to ionize, while the later infrared pulse will have its energy absorbed by the free electrons, causing the plasma to explode. Although the plasma is extremely hot, the temperature and energy aren't damaging in and of themselves—rather, the concussive force of the explosion creates enough of a pressure wave to knock targets off their feet, while the electromagnetic radiation affects their nerve cells and causes extreme amounts of pain. Just like the *Star Trek* phaser, this technology can stun targets on a low-power setting, but can easily kill them if sufficient energy is given to the device. Presently known prototypes can hit a target up to 2 kilometers away and have the potential to be scaled down from their present size—which must be vehicle-mounted—to the rifle-sized device envisioned by *Star Trek*.

Much more powerful than a handheld phaser, a phaser rifle was capable of causing greater amounts of damage and destruction than any other type of this weapon save those attached to a ship. They proved extremely useful in many combat and security situations.

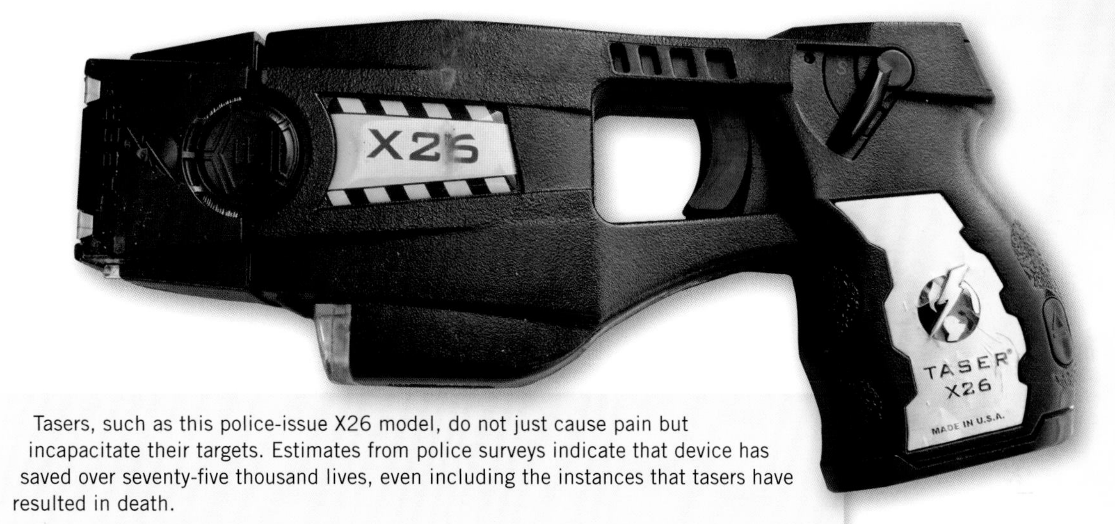

Tasers, such as this police-issue X26 model, do not just cause pain but incapacitate their targets. Estimates from police surveys indicate that device has saved over seventy-five thousand lives, even including the instances that tasers have resulted in death.

Besides directed-energy weapons, there's an alternative approach already in widespread use: electroshock weapons. Beginning in the 1930s with the electroshock glove, and arguably even earlier in the form of the cattle prod, scientists quickly realized that applying a high-voltage, low-current shock to the human body would override the mechanisms that normally control the body's muscles. Instead, the muscles spasm or twitch, rendering the target incapacitated. Known as a "stun gun," the most conventional application is the Taser, which was developed from 1969 to 1974 by a NASA researcher named Jack Cover. While the initial device, known as "Thomas A. Swift's electric rifle" (whose initials are the origin of the brand name Taser), contained gunpowder and was therefore classified as a firearm, modern versions either use compressed gas to fire darts with conductive wires attached or activate upon direct contact of the two conductive nodes with the target. In either case, a swift and powerful alternating current is applied, incapacitating whatever human being is targeted in a nonlethal fashion.

In *Star Trek*, phasers could be equipped with a number of custom settings for various tasks, including sweeps, spreads, or directed cutting through a wall, door or even a ship's hull. Those are some of the technology's applications that require far more power than we can implement at present.

While no one would mistake a Taser for a phaser, a new technology that combines the electrically conductive current aspect of the Taser with lasers and plasmas might be our best bet for bringing *Star Trek*'s device to fruition. Known as the electrolaser, this concept works by sending a laser through a medium such as air, which ionizes the electrons in it, creating a plasma in the shape of a channel. Then, a fraction of a second later, an electric current is sent through that plasma channel, achieving high voltage through a series of step-up transformers. The net result is a long-distance weapon that functions with the same properties as a contact Taser, with similarities to lightning. Because of its reliance on a conductive medium—the plasma in air—this version of a phaser wouldn't work in the vacuum of space, but it could be an incredibly effective distance energy weapon so long as there is a clear path between the holder of the phaser and the desired target.

Remarkably, current technology is quite capable of engaging in the phaser's most widespread technological use: the disabling of a human or humanoid adversary without long-term physical harm. Whether through a pulsed-energy projectile or an electroshock device, the human body is highly susceptible to being physically incapacitated through a number of nonlethal means. Unlike *Star Trek*'s phasers, however, these contemporary weapons have

Phasers can be used in a variety of combat situations.

limited utility when it comes to some of the phaser's other potential applications, like blasting through inanimate objects or cutting through high-density materials. To create a phaser-like device of that magnitude would require a laser-type setup with far more power than anything we've developed, as the most powerful laser system created to date—the 192-laser array implemented at the National Ignition Facility—delivers less than 10 kilojoules per laser, which is about the equivalent of a paltry two grams of dynamite.

Nevertheless, the fact that we've made so much progress in the time since a phaser was first envisioned should make us extremely optimistic about the future development of energy-directed devices. They're already far more safe and effective than rubber bullets and other nonlethal uses of force in law enforcement and riot control, and with time, we can expect these technologies to continue to improve. The next big step toward a *Star Trek*–style phaser would be to implement a mechanism that doesn't rely on a medium like air, or conductive contact like a Taser, to direct the energy. If we can overcome that next barrier, energy-directed weapons for use in a space battle, far beyond the capabilities of mere lasers, just might become a part of our reality.

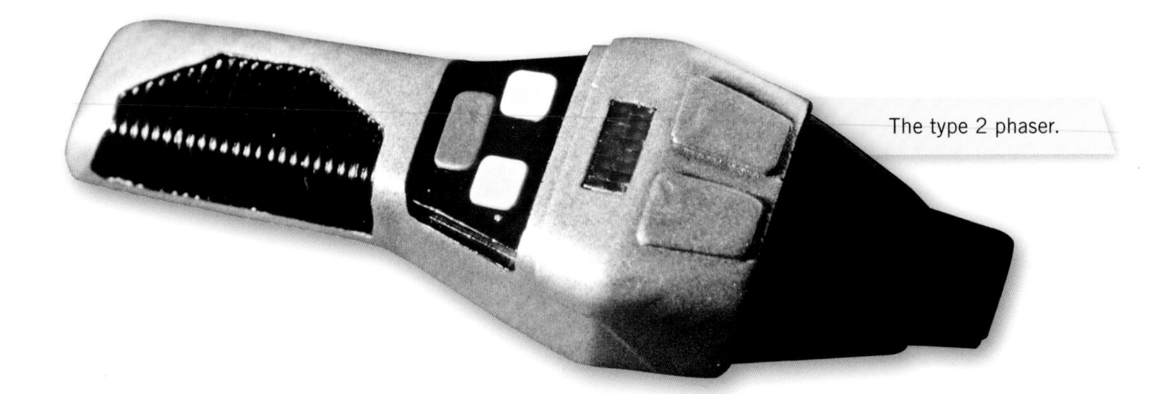

The type 2 phaser.

PHOTON TORPEDOES

The photon torpedo was the most powerful weapon aboard a starship from the twenty-second through late twenty-fourth centuries. Unlike a directed-energy weapon like a phaser, a photon torpedo represents the ultimate evolution of a projectile or missile-like object. Whether fired from a launcher toward a target, left like a mine in interplanetary or interstellar space, beamed to a location by transporter, or detonated on impact or otherwise, its explosive force was unparalleled as far as destructive power goes.

Among the most destructive armaments in all of *Star Trek*, the photon torpedo was instrumental in both attack and defense in firefights in space.

The *U.S.S. Enterprise*-D destroyed a number of ships with a direct hit from a single photon torpedo, including the *U.S.S. Lantree*, which met its demise in 2365.

A single direct hit from a powerful enough torpedo could obliterate not just a starship, but even an entire starbase or command center, particularly if shield capabilities were negligible.

Perhaps most memorably, Spock was buried in a photon torpedo casing that served as his coffin after his demise in *The Wrath of Khan*.

When the *Enterprise* NX-01 first encountered Klingons in the mid-twenty-second century, they found themselves far outgunned, as spatial torpedoes—containing conventional nuclear warheads—were no match for the much more powerful photon torpedo. By time the *U.S.S. Kelvin* was attacked in 2233, however, it was equipped with a full complement of photon torpedoes, although they were woefully ineffective against the late twenty-fourth century technology they were up against. A powerful enough deflector shield could stop a photon torpedo blast, but if the shields were taken down, a direct hit from a photon torpedo was among the most catastrophically destructive effects seen in all of *Star Trek*. There are even hints that a properly modified photon torpedo could destroy a small planet, although thankfully, no such wanton destruction was ever implemented to our knowledge.

The "torpedo" part of a photon torpedo is simplicity itself, appearing no more sophisticated than a standard metal pod. Anything with which a small craft or missile could be outfitted—propulsion, shields, a cloaking device, or even warp capabilities—can be used to outfit a photon torpedo as well.

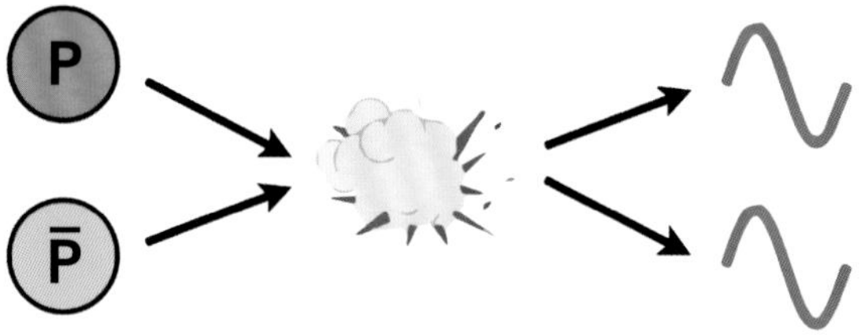

If equal amounts of matter and antimatter are allowed to collide, they will produce photons in a nearly 100% efficient reaction. Only the particles and antiparticles that cannot find one another during the explosion will survive.

Considering that payload-removed photon torpedoes have been modified and used for coffins, for smuggling crew members, for illumination, or even for exposing cloaked ships, it stands to reason that creating the actual matter-based torpedo poses no problems for even twenty-first-century science.The true scientific challenge is addressing the issue of tremendous destructive power. Conventional chemical weapons like gunpowder, TNT, or C-4 explosives rely on electron transitions within atoms to derive their power via energy release, typically converting less than 0.001% of their mass into energy via Einstein's $E = mc^2$. Nuclear weapons, either through splitting heavy atoms (fission) or combining light ones into heavier ones (fusion), are much more efficient, but still convert only between 0.1% and 1% of their mass into energy via $E = mc^2$. The ultimate in imaginable efficiency is matter-antimatter annihilation, which would convert 100% of its mass into pure energy. The "photon" in photon torpedo would arise from perfectly efficient matter-antimatter annihilation, a theoretical possibility that was demonstrated experimentally by twentieth-century particle physicists.

If antimatter containment (see page 43) were ever attained, then creating a photon torpedo would be as simple as constructing a torpedo case capable of containing equal amounts of matter and confined antimatter within it (appropriately separated, of course) and then allowing them to detonate with one another at the appropriate moment—either on impact or when they've reached a particular location. The most powerful nuclear weapon ever detonated in human history was the Tsar Bomba, a hydrogen bomb tested by the Soviet Union in 1961, which achieved a yield equivalent to about 50 megatons of TNT—the release of 2.1×10^{17} joules of energy. To equal that would require a reaction between just 1.2 kilograms each of matter and antimatter. But torpedo casings are much larger than a few kilograms of material: they take up the volume of a large filing cabinet.

Assuming that your matter and antimatter stashes were of approximately the same density as rock, a reasonably dense material, you could fit approximately 3 metric tons (3,000 kilograms) each of matter and antimatter in this space, which would give rise to an explosion of 5.4×10^{20} joules of energy—2,600 times as powerful as the Tsar Bomba, with a yield of about 130 *gigatons* of TNT. And if your dream was to destroy a planet, a properly rigged photon torpedo might actually get you there. Consider that all the planets we know of are made out of matter, so a 100% antimatter-loaded photon torpedo wouldn't have to bring its own matter to annihilate with; it could simply annihilate with the target.

The first two-stage thermonuclear device, the hydrogen bomb, was detonated in 1952 and was code-named Ivy Mike. The explosive yield was the equivalent of 10 megatons of TNT, about 20% the energy of the Tsar Bomba.

At maximum yield, a properly detonated configuration of photon torpedoes could bombard a target with arbitrarily high amounts of energy as well as alter its momentum, which proved very useful in an encounter with a soliton wave in *The Next Generation*.

The gravitational potential energy binding Earth together is 2.24×10^{32} joules, or about 400 billion times as much energy as our photon torpedo, maximally filled with matter and antimatter, could release. The size of the torpedo need not be increased, but perhaps the density of the antimatter inside could be: instead of rock-density material, what if neutron-star-density material were inside? It would have to be made of antineutrons rather than neutrons, but a large-apple-sized sphere's worth of neutron star material—about a 7-centimeter-diameter sphere—would be enough to do it. Although the surface neutrons would decay due to their instability, the antineutron half-life of fifteen minutes ensures that the vast majority of the antimatter warhead would make it to its desired destination. If you beamed the antimatter-laden torpedo into the planet's interior, even better; not only would you avoid the problem of the tremendous recoil you'd experience firing such a massive torpedo, but there'd be no way out for the antimatter except to annihilate with the interior matter, blowing the planet apart.

Practically speaking, we're a long way away from developing a photon torpedo, primarily because our antimatter storage capabilities are so rudimentary compared to what's needed. While we've made tremendous advances in recent years to create neutral antiatoms and confine them for minutes at a time, we have yet to create enough antimatter in total to create even one weapon that rivals atomic bombs in energy. That's the main limiting factor, however—in principle, we know how the physics of this work, and we've experimentally created all the various components necessary. All that's missing is the practical technology to bring the photon torpedo to fruition.

Just like in *Star Trek*, it should be no problem to have a torpedo with various yield settings, simply by varying either the amount of antimatter, the configuration of the annihilation pattern, or the amount that actually detonates. Any technologies otherwise available in the *Star Trek* universe—like shields, cloaking, warp drive, or transport ability—could be applied to photon torpedoes, making them even deadlier weapons. While many technical obstacles need to be overcome, there are no scientific dealbreakers standing in the way of the photon torpedo; it's essentially just a question of our created technologies catching up with what we know is possible.

COMMU

CATIONS

When *Star Trek* first premiered in 1966, long-distance communications by direct dialing telephone calls (as opposed to calls connected by switchboard operators) were still relatively new, and not in use everywhere. In many places, human operators were still needed to connect phone calls. The copper wire through which the signals were transmitted ensured a tinny, static-filled quality overlaid on top of everyone's voice, and there was only one massive handset and transmitter combination available: the one the phone company provided you with and installed. Yet there, on television, was the realization of a very different future—one where communications could take place virtually instantaneously across tremendous, even interplanetary distances, through a tiny, wireless device with no external power source. The dream of instantaneous, wireless communications set the public's imagination aflame.

Star Trek took that dream further, envisioning instantaneous transmission across the galaxy through subspace, while universal translators would enable people who spoke different languages—including humanoids from entirely different species—to understand one another completely in real time. In an era when, in the real world, relations between foreign countries on the same planet would often break down due to the language barrier, this was an optimistic view of how peaceful the future could be, even when the Federation was up against a potentially hostile, warlike race.

Communicator badges, envisioned for the twenty-fourth-century versions of the show, further streamlined communication. *Star Trek* was directly responsible for inspiring many of the technologies we now take for granted, including many of the devices we now use on a daily basis. Today, spoken-voice commands and potentially, in the future, thoughts will enable us to control who we communicate with and when, anywhere in the world, virtually instantaneously. Earth is more connected than ever and is only becoming more so, and for a great

Thanks to the power of subspace communications, starships could communicate with bases, commanders, or other worlds without a noticeable time delay of any

SUBSPACE COMMUNICATION

"Hail that ship," the captain bellows, and a crew member sends a signal out into space, where the other starship receives it in practically an instant. Even though the hailed vessel may be millions of kilometers away or more, there's not even a second's worth of delay in transmissions. Near-instantaneous communication at these distances requires circumventing Einstein's theory of special relativity, and so in the world of *Star Trek*, light speed simply won't do. Instead of communicating with radio (or other electromagnetic) signals sent conventionally through space, a new type of communications—through *subspace*—would need to be invented. Suddenly, not only can you avoid any delay of minutes or even seconds to speak with a ship at interplanetary distances, but with enough signal power, you can communicate with allies or adversaries light-years away in near-real-time.

As early as the twenty-second century, subspace communication in *Star Trek* was used for sending and receiving messages over large distances, with the "ancient" technologies of antennae and amplifiers being necessary for any appreciable distances to be traversed. The bandwidths of these early communications were small as well, with the postwar Earth-Romulan treaty established entirely over subspace, by radio alone. By the twenty-third century, communications officers weren't merely translators and facilitators but also controlled a ship's subspace communications system and maintained the subspace logs. Messages were sent between planets and starships as rapidly as one might send an email or text message or make a call today, with even less delay than we experience from one side of Earth to the other. And by time the twenty-fourth century came around, a galactic relay network made video transmissions across many light-years possible, again at near instantaneous speeds. Although communication was very, very fast, it wasn't quite instantaneous—when the crew of the *Enterprise*-D were transported to the edge of the known universe, they found it would take over fifty years for a subspace signal to reach Federation space.

Unfortunately, there is no such thing as subspace in the real world, but that doesn't necessarily mean that a technology like this is physically impossible. Quite to the contrary, of the two simultaneous

The *Enterprise* NX-01 deployed three subspace amplifiers in the mid-twenty-second century, two of which were destroyed by an unknown alien vessel.

innovations required—a way to communicate faster than light and a way to collimate signal power more effectively—one may be on the horizon and the other is already in the process of becoming possible. Although faster-than-light communication is impossible under Einstein's theory of special relativity, the same potential physics that underlies warp drives (see page 10) could lay the foundation for an alternative approach to this technology. It may even prove to be easier, as instead of needing to transport an entire massive spacecraft through a warp bubble, it would only need to send a radiation signal. This means that a much smaller region of space would need to be transformed into the Alcubierre spacetime, and that simply controlling where the warp field began and ended would enable an arbitrarily fast signal speed. Much like sending a starship at warp speeds, the physics behind creating this type of spacetime would be the same—requiring negative mass or some other form of negative energy—but it isn't ruled out from models involving exotic physics.

A subspace relay network was set up some time between the twenty-third and twenty-fourth centuries, replacing the use of subspace amplifiers and enabling transmissions to take place across large interstellar distances. Relay stations, therefore, were important from a strategic and military standpoint, as well as for simple civilian uses.

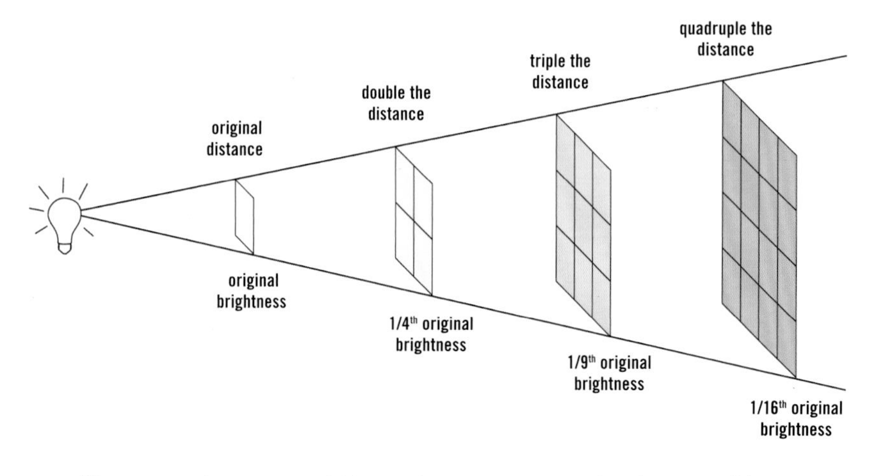

The second component that subspace communications would need, however, is the ability to send a signal across long distances that wouldn't degrade the same way conventional signals do. Radio waves, for example, see their energy density spread out according to an inverse-square law ($1/r^2$, where *r* is the radius—the distance from the source of the signal), meaning that transmission to a distance ten times as great requires a hundred times the power. The reason the signal falls off this way is because electric charge is conserved, but there are two types of charge, positive and negative, and charges are free to move. The physical intricacies of this mean that you get what's known as dipole radiation, which is why the signal drops as $1/r^2$; this affects and limits all electromagnetic signals.

ABOVE: Any electromagnetic signal, including radio transmissions, will lose energy density as it spreads out through space at the speed of light. A signal with four times the magnitude is necessary for reaching twice as far.

OPPOSITE: An illustration of the gravitational waves emitted from merging black holes, from a simulation at the Albert Einstein Institute in Germany.

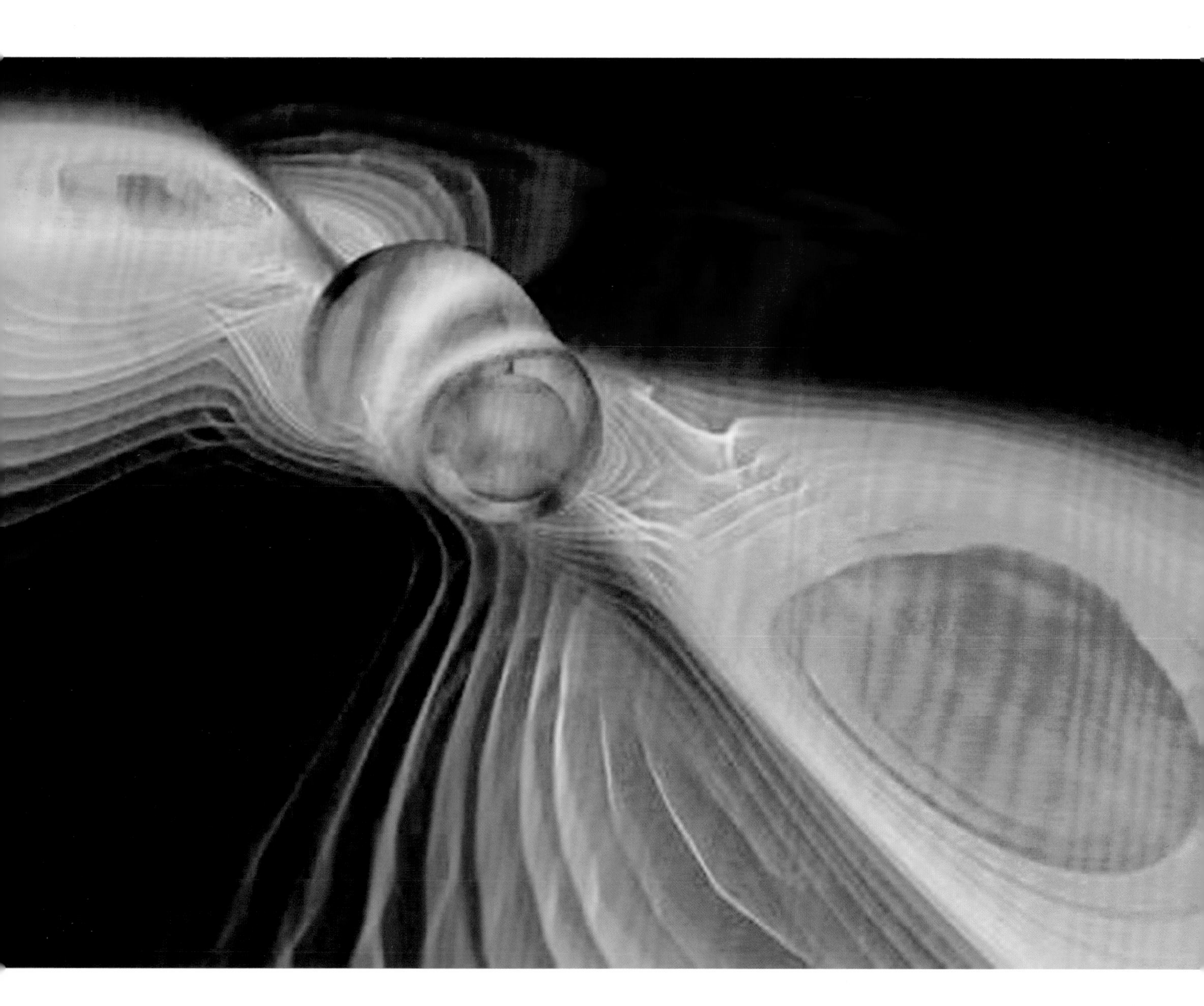

For gravitational fields, however, not only is mass conserved just as electric charge is, but there's only one type of "charge," known as mass (positive only). The one type of charge, along with momentum conservation, makes dipole radiation impossible. Instead, you need a severe time variation in the distribution of the mass system, or the system's moment of inertia, in order to create gravitational radiation. Under those specific circumstances, this creates *quadrupolar* radiation, which only falls off as $1/r$, not $1/r^2$. The first direct detections of gravitational waves by the National Science Foundation's Laser Interferometer Gravitational-Wave Observatory (LIGO) in 2015 validated the existence and form of gravitational radiation, confirming its quadrupolar nature and the fact that it falls off as the inverse of the distance, not the inverse of the distance squared.

By combining warp-field technologies that enable the transmission of matter and/or radiation through space at speeds far exceeding the speed of light, and controlled gravitational radiation, which falls off far more slowly with distance than conventional electromagnetic radiation, faster-than-light communication across incredible distances may actually be possible. While subspace communication as depicted in *Star Trek* may not exist, owing to the fact that subspace itself doesn't exist, achieving the same practical result—sending information across the universe to a distant recipient—may be plausible after all. At this point in time, the amount of energy required to create a detectable gravitational wave signal is tremendous, as is the precision required to detect it on the other end. But the ability to send any signal across space at faster-than-light speeds was thought to be complete fiction when *Star Trek* first premiered, hence the invention of an entirely new type of space to explain it.

Current technology has not yet reached the stage where faster-than-light transport—of matter, radiation, or information of any type—has been accomplished, but the detection of gravitational waves is one of humanity's most recent, exciting scientific advances. While the first detections will focus on the strongest signals—such as stellar-mass black hole mergers, neutron star collisions, supernovae, and objects spiraling into supermassive black holes—this class of technology has already improved by many orders of magnitude over the past two decades, and will likely continue to do so well into the future. As we progress from ground-based to space-based gravitational wave detection—like the Laser Interferometer Space Antenna (LISA) and, potentially, an array of laser interferometers like NASA's proposed Big Bang Observer (BBO)—gravitational radiation signals of a wider range of frequencies and of much smaller power levels will be detectable, and the study and application of such waves will be explored in earnest. While it's very difficult to predict what new technologies will emerge from this, gravitational waves could prove to be the ultimate long-distance communications tool. Once they are fully understood, generating them and transporting them with additional boosts through space just might lead to the ultimate realization of instantaneous interstellar communications!

Free-floating lasers at great distances from one another in space should enable the detections of gravitational waves to far greater sensitivities and over a much wider range of wavelengths than anything achievable on Earth.

Touch your badge, say the name of the recipient, and communication is imminent. Real-life technology has advanced so far, so rapidly, that this now almost seems antiquated.

COMMUNICATORS

Kirk flips open his communicator, listens for the familiar *chirp*, and whether he's on another ship or down on a planet, a simple "Kirk to *Enterprise*" will put him in touch with whomever he needs.

We have been contending with the problem of long-range communication for as long as humans have been around. Once your desired recipient was out of earshot and out of sight, sending a message or even alerting someone to your presence often took a herculean effort. For millennia, visual signs like beacons or smoke signals were the only alternative to sending a messenger. In the nineteenth century, the invention of the telegraph allowed the first rapid, long-distance communications, so long as you had a wire or cable connecting the source to the destination. The telephone followed suit, but with the same limitation. In the twentieth century, wireless communication became possible, first with the invention of the radio and then with the mobile two-way radio. This was further adapted into a walkie-talkie (or handheld transceiver), requiring an attached bulky, mobile apparatus in order to use it. But the imaginings of *Star Trek* in the 1960s, at a time when telephone signals transmitted through a network of copper wires were state-of-the-art, the idea of long-distance, wireless two-way communication through a tiny, handheld device was unheard of.

By using a series of touch communicators throughout the ship, it became easy for anyone in the twenty-third century to call anyone else at will. Mobile, handheld flip communicators enabled landing parties to stay in contact from great distances apart, even calling the ship with no problem from hundreds or even thousands of miles away. By time the twenty-fourth century came around, the handheld communicators had been replaced by simple comm badges, emitting that familiar *chirp* when touched and fully capable of reaching anyone within range.

Signaling for help in a remote location used to be a fruitless endeavor. With the advent of a communicator, a call for help was almost always heard.

It might seem quaint by today's standards, considering the power and the myriad of capabilities contained in a modern-day smartphone, but the original handheld communicators were a revolutionary idea. In fact, this is one example where a technology imagined on *Star Trek* may have directly inspired an inventor who would model a new technology precisely on an idea born purely of science fiction. In 1973, inventor Marty Cooper conceived of a handheld, long-range mobile phone, completely independent of any external power source like a car, truck, or backpack. Over the next decade, working at Motorola, he developed and brought to market the first cellular telephone in 1983, even making the very first such call himself.

It took a number of big innovations to make a device like this possible here on Earth:

- the miniaturization of electronics and batteries, enabling a single, compact, low-power device to house a completely working product capable of receiving and transmitting audible sounds wirelessly;
- a cleverly-sized and attenuated antenna, capable of receiving and transmitting signals at the appropriate frequencies for communication;
- a network of high-powered transmission towers on the ground, capable of receiving and broadcasting signals from and to individual, handheld devices;
- and a slew of satellites in space that send signals to and from those transmission towers.

As the communications officer, it made sense to have an earpiece directly connecting you to whatever the person or alien on the other end had to say.

That's an awful lot of infrastructure, but that's what it takes to broadcast a signal of sufficient magnitude from such a small device.

Flip-phones, a design modeled on the original *Star Trek* communicator, became widespread in the late 1990s and 2000s, lasting until they were

A site-to-site transport, a combadge, and an auto-firing phaser added up to a very clever trick to fool a large contingent of the *Enterprise*-D's security team hellbent on capturing and brainwashing an elusive Wesley Crusher.

outpaced by brick-style smartphones in the 2010s. Many of today's smaller devices, such as wearables and smart watches, may be more reminiscent of twenty-fourth-century badge communicators. Additional innovations—such as subtle, nearly silent vibrations and earpieces—as well as text messaging and other modern applications, make communicating much more varied and inconspicuous. Perhaps the first in-ear Bluetooth devices were even inspired by Lieutenant Uhura's wireless communications earpiece?

Star Trek predicted numerous applications attached to mobile communications devices, like today's location and transponder technologies, including GPS, LoJack, and RFID. A strategically placed location tracking device can assist anyone attempting to find and capture (or worse) a particular target.

Practically instantaneous communication between any two terrestrial locations, limited only by the physics governing this universe, has been physically possible ever since humans first discovered and isolated electromagnetic radiation, uncovering the quantum nature of light. So long as the algorithm to encode and decode information is known (or decipherable) by both the sender and the receiver, a signal of a sufficient magnitude sent and received should enable communication across any distance. So long as your signal has enough strength to reach its target at a certain distance—where physics dictates that the signal power spreads out with the distance squared from the signal's source—you can reach anyone at the speed of light.

It's truly incredible that in an era where the only phones people routinely had were the hard-wired sets with rotary dials offered by the phone company, *Star Trek* was able to envision a future means of communication that literally inspired the technology that humans would go on to develop over the coming decades—only for fact to quite rapidly surpass fiction in many areas! What's still missing, however, is the ability to transmit any sort of information through a small device across long distances without a high-powered relay device like a transmission tower. Perhaps the necessity to broadcast through space—rather than through subspace—is the last piece of the puzzle preventing us from realizing the dream of low-power, long-range, instantaneous communication and location technologies.

Quantum beacons, used to detect objects invisible to conventional scanners, required tremendous power even in the twenty-second century.

UNIVERSAL TRANSLATORS

"*Heghlu'meH QaQ jajvam*!" cries the Klingon entering battle. While many fans will recognize the phrase as meaning "Today is a good day to die!," those of us who aren't fluent in Klingon—including many Starfleet crew members in the *Star Trek* universe—have to rely on the translations of others. This can be unreliable, subjective, laborious, and time-consuming, among many other potential flaws.

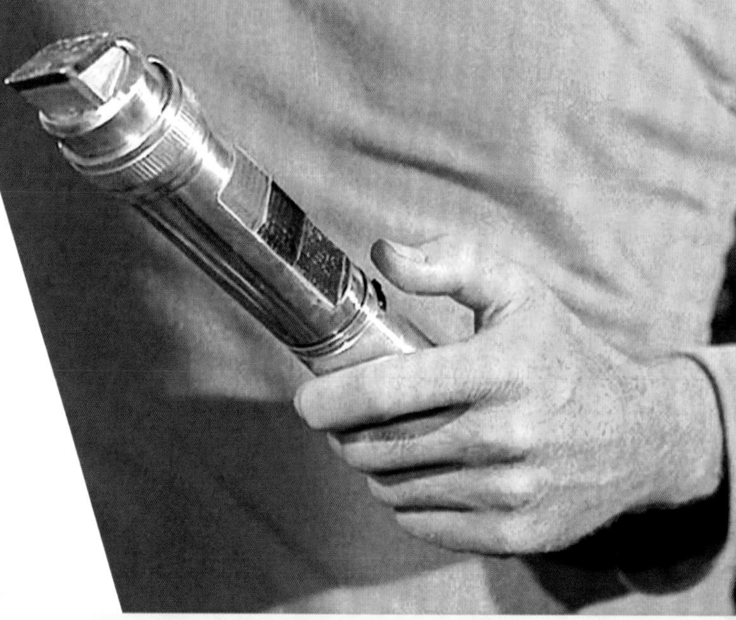

A landing party universal translator of the twenty-third century.

Attempts to get people to adopt a "universal language" here on Earth have proven very difficult, as at no point in history have a majority of the world's population been even marginally fluent or literate in the most widespread languages like French, English, or Mandarin. Even languages explicitly constructed to be universally spoken, like Esperanto, have failed miserably when it comes to adoption. But *Star Trek* envisioned an incredible alternative: rather than having citizens of different countries or planets learn one another's native language to communicate, what if there were a technology that automatically translated for them, on the fly, in real time? A universal translator device that could process a speaker's language and cause the full meaning to register in the listener's native language, without even the slightest delay, would break down practically all the communication barriers plaguing speakers of different languages in the world today, not to mention between the worlds beyond our own.

promise of encountering new, alien languages—and the e of understanding them, translating them, and facilitating nunication between new species—was the deciding factor tting Hoshi Sato to join the crew of the *Enterprise* NX-01.

Ideally, a universal translator would take whatever language was natively spoken and transform the words and phrases—nuances and all—into the listener's preferred language. If the spoken language were understood well enough, a translation matrix could be constructed, enabling practically instantaneous translation to a language that the listener could comprehend. Although these devices were available to all *Star Trek* crews, having been invented and widely distributed by the mid-twenty-second century, communications officers were required to be language specialists for quite a long time thereafter. Hoshi Sato's skills were particularly useful aboard the NX-01 for reading foreign languages and proved vital in the development of the linguacode translation matrix, which eventually enabled new languages to be translated practically instantaneously once enough data was gathered.

The actual universal translator was a small device that would either come with a keypad and display that attached to a communicator or clipped onto clothing as a standalone implement. In addition, known languages were built in to the communicators themselves, even as early as the 2150s. By the twenty-third century, universal translators were automatically built in to all Starfleet ships' and shuttlecrafts' communications systems. Other races communicating with a starship would no longer necessarily need explicit words and phrases translated, but rather their intentions, demeanor, cognitive patterns, and ideas could be scanned and pieced together into decipherable sentences. By modifying a universal translator, Kirk and Spock were able to communicate with the Companion through exactly that method. In the twenty-fourth century, even the comm badges that each crew member was outfitted with came with built-in universal translators capable of quickly learning and translating languages from an entirely new species, as with the nanites in *Star Trek: The Next Generation*.

The universal translator is a combination of two emerging technologies, neither of which was available when *Star Trek* was first conceived: natural language processing and an accurate translation program. In the 1960s, the only way to input information into a computer was manually, by a variety of encoded media (such as punch cards and disks) or by a command-line interface. The concept of speech recognition was investigated at Bell Labs as early as 1932 but was abandoned as a pipe dream until Raj Reddy—a Turing Award–winning computer scientist—designed an algorithm to recognize and issue spoken commands for a game of chess. During the late twentieth and early twenty-first centuries, DARPA-funded competitions resulted in the best speech recognition and language processing to date. The first commercially successful speech-recognition technologies were brought to market in the 1990s, each with a vocabulary larger than the average human's, ushering in an era in which this technology would become ubiquitous. With the debut of Apple's digital technology assistant, Siri, in 2011, speech recognition came into the mainstream. Today, other digital assistants such as Cortana and Alexa, as well as Google's speech-recognition software (and Google Now), are available to anyone with a smartphone, with more than thirty compatible languages supported. (Although strong accents are still a maddening but often-hilarious problem.)

An accurate translation program is the other major component necessary for this technology to be brought to fruition. While the most common modern translation programs out there—programs like Babelfish, Google Translate, or Bing Translate—often produce laughably awkward and inaccurate results, they also represent huge improvements over similar programs from just fifteen to twenty years ago. (Want a fun test of how good they are? Input a modest English sentence, translate it into another language, then copy and paste that result into the input box and translate it back into English. You might be surprised at how different your end product is from your initial sentence!) In general, translation software algorithms are now good enough that one of these programs can preserve roughly 75 to 90% of the original meaning when brought into another language, although many of the idioms and nuances require more careful programming and attention than direct word-to-word translations. Premium software packages such as Babylon 10, Power Translator, and Promt can often do better, supporting up to seventy-seven languages with features such as the capability to automatically recognize the spoken language, do text-to-speech, and give context-specific translations. With the incorporation of machine learning algorithms into the translation software, translations to and from context-driven languages continue to improve dramatically. Some programs are even capable of scanning and recognizing written text that appears in images, with apps like Word Lens—whose technology was acquired by Google and is now available as part of Google Translate—capable of recognizing a few languages in real time and translating them into your preferred language!

But as *Star Trek* envisioned it, a true universal translator would enable real-time conversations between two people who didn't speak one another's language. Moreover, one user's spoken language would be translated so that the other user would hear that language in their own tongue, with the original meaning preserved. The advent of machine learning has meant that the more a variety of native speakers provide phrases, context, meaning, and feedback to a service, the better it learns.

Even noncarbon-based life that had developed their own language would be communicable with a universal translator, as the crew of the *Enterprise*-D found out when they were (correctly) identified as "ugly giant bags of mostly water" by an intelligent crystalline lifeform.

Even a completely successful universal translator would have its own limitations if a previously undiscovered or unprogrammed way of communication were encountered. "Shaka, when the walls fell," or a failure to communicate, was exactly the situation the Federation found themselves in when faced with the Tamarians, who used language entirely differently from all other previously known cultures.

In December 2014, a preview of Skype Translator became publicly available, with speech recognition, statistical machine translation, and computer speech synthesis providing the beginnings of true real-time translation. While only eight languages are presently supported, October 2015 saw the integration of Skype Translator into Skype for Windows desktop, with the interface becoming publicly available in 2016. But perhaps the biggest advance has come courtesy of Waverly Labs, who in May 2016 introduced an earpiece and app combination that enabled direct, near-real-time translation from any two supported languages between speaker and listener, so long as they both have earpieces equipped. It's quite foreseeable that in the very near future, translation devices will become so small and powerful that a single earpiece could receive, interpret, translate, and synthesize speech for the wearer all on its own.

As with all machine-learning endeavors, experience is key. We can only expect a universal translator to be as universal as the frequency with which a language has been studied and understood. Just as direct translations can be technically correct on a word-to-word basis but may lose meaning when phrases, idioms, metaphors, or sentences are considered, so will a universal translator fail until the algorithm powering it "learns" these nuances. "Shaka, when the walls fell," might be easy enough to translate literally, but understanding that it means "failure to communicate" is a whole different matter. Yet as the structure, meaning, and expressiveness of a language's speakers becomes better understood, the translation becomes progressively more and more natural-sounding to the listener, regardless of the listener's language.

While the current incarnations of universal-translator technology can only translate known languages—and even then, only the languages a device has been programmed to translate—this technology has already nearly reached the capabilities envisioned by *Star Trek*, hundreds of years ahead of schedule. While someone's neural patterns aren't quite read directly, their sentences can be listened to and processed for meaning at the speed at which their language permits, and then translated for the listener's benefit with only a few seconds of delay. The only limitations a modern human using a perfected version of this technology would encounter would be the same limitations a Starfleet crew member might encounter: until the nuances of an unknown language-speaker's usage are understood, a word-for-word, dictionary-level translation might not facilitate communication as much as we'd want it to!

GALAXY CLASS STARSHIP
USS ENTERPRISE • NCC-1701D

COMPUTING

In 1966, when *Star Trek* was first introduced to the world, computers were gigantic machines that were only just beginning to gain mainstream acceptance. The most powerful computer at the time was Control Data Corporation's CDC 6600, which was capable of up to 3 megaflops, or 3 million FLoating-point Operations Per Second. Only CERN and Lawrence Berkeley Labs had one of these machines at the time *Star Trek* first aired, and the show's visions for what computers would become were completely revolutionary.

A ship's computer was imagined that not only was more powerful than any human or electronic device of the day, but could speak and receive oral commands and translate them directly into whatever novel computer code needed to be written. Inconceivable amounts of data could be taken, recorded, and stored concerning stars, planets, life, diseases, and anything else that was encountered. Computers would be handheld and have visual, interactive interfaces. And by the time *Star Trek: The Next Generation* came along, artificially intelligent humanoid androids, holograms, and entire lifelike personalities and intelligences could be stored on a single stick no bigger than two fingers.

In many ways, the computational power envisioned by *Star Trek* was very forward-looking, but in some ways, it didn't go nearly far enough. Fifty years after *Star Trek* first premiered, the single computer record for computations reached 93 petaflops, or more than 30 billion times faster than 1966's CDC 6600. As computing power continues to rise at an unabated rate, the most difficult computational problems continue to fall to our advances, from mathematical proofs to complex games like Othello, chess, and go to predictive modeling of complex, chaotic systems. While *Star Trek* dreamed up a great many applications that we're still

Capable of storing so much data that an entirely new scale was invented—quads instead of bits—the isolinear chip represented a vision of data storage that didn't involve magnetic tape or moving parts.

ISOLINEAR CHIPS

When you're traveling through the galaxy at warp speed, scanning every cubic centimeter of space for whatever forms or patterns of matter and energy might be present, the amount of information you need to store starts to add up, and *fast*. To program something as advanced as a warp engine, a runabout, or a PADD (see page 127) or to simply store incredible amounts of audio, video, and other forms of data, you need a novel, efficient technology leaps and bounds beyond anything that was known in the twentieth century. The technological solution to this problem, presented in the first season of *Star Trek: The Next Generation*, was an optical storage device unlike anything known at the time: the isolinear chip.

The duotronic computer revolution of the twenty-third century meant that unprecedented amounts of data could be stored in either binary or trinary formats. Duotronic enhancers were the greatest storage devices until the early twenty-fourth century, when the more efficient isolinear chips replaced them. Measuring data in units of *quads*, and used in conjunction with nanotech

The isolinear chip, right, replaced the older data card, left. Optical data storage made its public debut in *Star Trek: The Next Generation*.

By contracting all of space down to two dimensions on his one-of-a-kind warp 10 flight, Tom Paris and the *Cochrane* gathered more data in less time than any ship before it.

processors, a single standard chip could hold up to 2.15 kiloquads of data. While the information encoding a complete human being might encompass teraquads worth of data, the Emergency Medical Hologram program took up 50 petaquads of storage capacity. The largest accumulation of data in the Federation came during Tom Paris's warp 10 test flight, in which the shuttlecraft *Cochrane* traveled at infinite speed across the universe and collected data on the entire sector the *U.S.S. Voyager* was in, totaling over five exaquads of information. As detailed in *Star Trek: Voyager*, Borg technology may have been able to gather and store even more data than that.

At a time when the sum total of every computer system on Earth was less powerful than a single one of today's entry-level smartphones, it's no wonder that Kirk's *Enterprise* was a little light on the details of data storage and computing power. By time *Star Trek: The Next Generation* premiered twenty years later, though, computational power and media in general had evolved tremendously. Individual VHS tapes could store up to six hours of video; hard disk drives had reached capacities of tens of megabytes; tape backup units, although slower, could store even greater amounts of data. But *Star Trek* still needed to give us something to dream about, something that far surpassed the technology of the time. Enter a new type of technology: optical data storage, written in a new unit of data measurement (quads) and kept on an isolinear chip.

Optical storage devices weren't just a fanciful string of words at the outset of *Star Trek: The Next Generation*. It was a developing technology that became widespread in the 1990s with the development of compact discs and then CD-ROMs, DVDs, and DVD-ROMs, and it still persists with modern day Blu-rays and video games. There are many ways to store data, all involving some storage format to encode information at a fundamental level (with at least two options), in a way that can be as densely packed as possible and that can be retrieved, decoded, and put to use.

In a book, the alphabet is the storage format, the lettering and pages pack the information densely, and the book and human eyes are how the information is stored and then retrieved. For optical storage, the storage format is typically a "pit" in the material (where a pit is a 1 in binary language and the lack of a pit is a 0), the density is defined by how finely and precisely as you can etch your pits into a smooth surface, and the retrieval and decoding is done with a laser, which reads the depth of the rapidly-spinning surface by measuring the reflected laser light's light-travel time. A mid-1990s CD-ROM could hold the same information as roughly 1,100 books written in English; today's top-of-the-line Blu-ray disc can hold over 200,000 books—or one copy of *Star Trek Beyond*.

The rapid advances in the storage capacities of these devices have come due to the decreases in separation to which we can etch bits (from microns down to nanometers), the finer wavelength of the laser (Blu-rays use a 405-nanometer laser, as opposed to the nearly 700-nanometer wavelength used by CD-ROMs), and the increased densities and number of layers that can be encoded on a disc. Potentially, an extra improvement can be added based on either an angle of reflectivity or a varying depth of the pit in the individual device, leading to more options than simply a 0 or 1 encoded on the surface. Perhaps *Star Trek*-style trinary encoding is actually on the horizon!

But optical devices are fundamentally limited—as are all methods of data storage—by these factors as well. Because the wavelength of laser light comes with a particular built-in distance scale, there is a finite limit to how shallow you can make your pits. Because the surfaces of optical devices need to be made out of actual materials, there's a limit to how densely you can pack the information together.

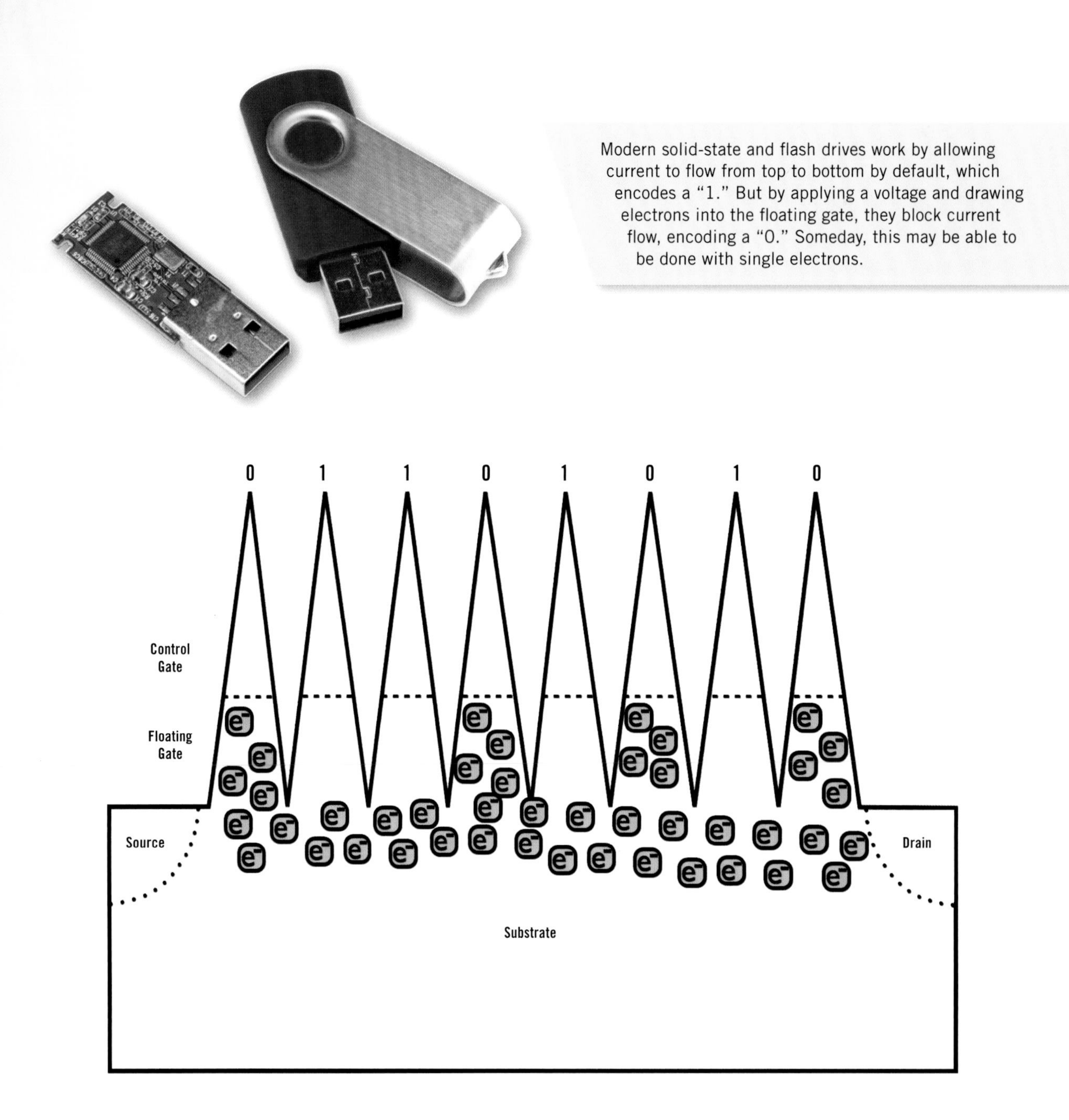

Modern solid-state and flash drives work by allowing current to flow from top to bottom by default, which encodes a "1." But by applying a voltage and drawing electrons into the floating gate, they block current flow, encoding a "0." Someday, this may be able to be done with single electrons.

Moore's law, the idea that circuits, storage, processors, and other metrics of computing power will continue to double in a given amount of time, is eventually going to run up against the quantum limit of matter itself. Today, optical devices compete with traditional hard drives and solid-state drives for packing the greatest amount of information into the smallest amount of space.

It's the solid-state drives that are the most similar to *Star Trek*'s envisioned isolinear chips. While optical and hard disks spin rapidly in order to be read (either optically or electromagnetically), solid-state drives store data on interconnected flash memory chips, and flash memory currently holds the record for most densely packed information structure. A single electron can be deposited in a specific quantum state (for a 1) or not (for a 0) in each memory cell, with cells approaching 10 nanometer spacing as of 2016. They are small, they do not scratch, they do not need to be spun in order to be read, they're durable, and they can hold tremendous amounts of data in tiny amounts of volume. In addition to solid-state drives, USB drives and memory cards are also examples of flash storage. In a device the size of your thumb, current technology can already hold 1 terabyte (10^{12} bytes) of data.

While there isn't necessarily a good way to convert between *Star Trek*'s fictional quads and the real world's bytes, the isolinear chip did a remarkable job of predicting both optical storage devices and handheld, insertable chips that can store data and memory, and can run their program on any compatible device into which they were inserted. As miniaturization continues, and storage capacities increase in both traditional ways and through the development of new techniques, it's clear that the concept of the isolinear chip is here to stay. This is one case where not only did the science-fiction predictions of *Star Trek* become reality, but reality leapt past the twenty-fourth century more than three hundred years early.

Entire time periods—real or fictional—could be recreated in the holodeck, complete with character personalities, idioms, and affectations particular to the scene.

THE HOLODECK AND HOLOGRAMS

The grid-lined walls of the empty room you're standing in disappear, and you find yourself immersed in any environment you can imagine. Imaginary worlds, landscapes, vehicles, living creatures, and even human beings suddenly spring into existence, seemingly alive and sentient. You can interact with them as you would any other programmed being, but the big difference from what you might expect over a conventional hologram comes when your senses other than hearing and sight get involved. Simulated flowers and plants have scents; simulated food and drink can be tasted and consumed; simulated terrain, weapons, and even creatures can be touched, including creatures as complex and intricate as a human being. And when you do touch them, depending on what type of program you're running, you're free to engage in any activity you can imagine, from adventure to combat to romance. Although this could easily improve human life by creating training simulations or exercise environments that would otherwise be considered unsafe, or allow the recreation or simulation of a situation that requires advanced analysis, by far the most widespread application would be for entertainment.

The holodeck enables people to indulge their fantasies, sometimes to the exclusion of their real-world responsibilities in a phenomenon known as holodiction.

"Computer, begin program."

The U.S. military has already adopted virtual-reality tools, making use of wearable devices, as part of its modern-day training strategies.

Holodeck technology wasn't a routine part of the original *Star Trek* series, but at some point in the twenty-fourth century it became widely available. Crew members of the *Enterprise*-D made liberal use of it, most frequently as an escape from their day-to-day lives aboard the ship. While leisure activities varied tremendously depending on the user's preferences, romantic encounters and heroic fantasies were by far the most common programs run by the crew. Will Riker created an imaginary love interest for himself named Minuet; Worf had a particular calisthenics program in which he would engage in ferocious combat situations he couldn't practice anyplace else; Geordi La Forge created an imaginary version of scientist Leah Brahms to solve an engineering problem, but later wound up developing feelings for her; Data was able to arrange a poker game against Albert Einstein, Isaac Newton, and Stephen Hawking. Perhaps most memorably, Reginald Barclay concocted a series of fantasy scenarios in which he would physically overpower the senior crew—notably La Forge, Riker, and Picard—and then have more intimate encounters with the female crew members. And in arguably the seamiest of applications, Quark rented out the holosuites on Deep Space 9 for the sordid satisfaction of any of his patrons' sexual proclivities.

Some sensations, perhaps, won't be able to be simulated properly. For those, the proper resources—like water on *Star Trek*'s holodeck—will need to actually be brought in or replicated separately.

But the holodeck wasn't merely for diversion. Want to train an engineer on how to deal with a warp core breach without having to create an expensive, deadly scenario? Want to see if a potential ship captain has what it takes to send a crew member to their death for the good of the mission? Want to simulate starship battles or visualize otherworldly, even catastrophic, scenarios? The holodeck can make it all possible, with no real-world ramifications. Additionally, the holodeck also proved its usefulness in forensic investigation, as it could recreate a variety of scenarios involving real and alleged crimes, and its use enabled justice to prevail, clearing Will Riker in the murder case of Dr. Nel Apgar.

In order to deliver on the promise of creating a simulated, lifelike experience indistinguishable from reality, holodeck technology must be able to create not just a visual projection, but a true "matter hologram." While this technology arises in *Star Trek* from holoemitters, force fields, and holomatter, there is no real-world analog as of yet. Holograms are simply projections of light that can either be a three-dimensional light field encoded onto a two-dimensional surface, like the security hologram on a credit card, or a reflection of an image from multiple directions, simultaneously, giving the appearance of a three-dimensional object. Although there are many illusions that appear to put a real, solid object (like your hand) in the same location as the hologram, there's nothing solid to interact with.

The goal of a lifelike, tangible hologram is to create a visual representation that exists in some three-dimensional space that can provide feedback to all of your senses. Since the 2000s, technology has begun to catch up to our imaginations. Three different effects are now coming together to make these tangible holograms real:

1. concave mirrors can create an apparent "projected" object to an observer from a particular set of angles,
2. position and motion sensors can detect where a person is located and communicate with the projectors, creating a uniquely interactive experience, and
3. the experience can be augmented with an ultrasonic force emitter that creates the sensation of pressure, contact, and touch.

This incarnation of "touchable holography" was developed at the University of Tokyo in 2009, and was the first major step toward a real-life holodeck.

Another holodeck-like experience that has been recently explored includes incredibly fast, pulsed lasers that ionize a very precise position in space, creating a temporary plasma. That plasma not only gives tactile feedback while being safe to touch (thanks to the incredibly fast speed of the lasers), but creates real photons visible to anyone present regardless of their perspective or position. Known colloquially as "fairy lights," they generate dots at a chosen point in three-dimensional space, where the number of pulses per second is determined by the speed, power, and other properties of the laser itself. As of 2015, this technology can produce between 4,000 and 200,000 pulses per second, with

With technology like the holodeck, the possibilities are endless.

each pulse having between 50 microjoules (for the fastest pulses) and 7 millijoules (for the slowest) of energy.

But perhaps the most holodeck-like incarnation is a virtual reality room, where an omnidirectional treadmill serves as the floor and the dome-shaped room is covered in projectors. Although no sense of touch currently accompanies it, the ability to move through a seemingly endless digital world projected everywhere within a human being's field of view simultaneously is one of the ultimate dreams of gamers everywhere. Instead of being glued to a single TV screen, you could instead be surrounded by the sights and sounds of an entire virtual world, with a single, dedicated room serving as a real-world holodeck. A competing technology is a set of virtual-reality wearables, which could give wearers not only the ability to see, hear, and potentially touch the virtual world around them, but to interact with one another simultaneously within that world. Much like in *Star Trek*, this wouldn't be exclusively for entertainment purposes and would lend itself to a variety of training and simulation scenarios, particularly of the dangerous variety. This technology could become widespread and commercially available as early as the mid-2020s.

While the ability to simulate a human being is still a long way off, as a machine that passes a robust Turing test—one of the best-known tests of whether a machine possesses humanlike intelligence—has yet to be developed, creating a virtual environment that provides anything from lighthearted entertainment to simulated life-or-death scenarios is well within the reach of modern technology. A variety of approaches exist to not only provide the visual and auditory sensations of being immersed in a virtual world, but tactile feedback as well. Certain sensations, however, much like in *Star Trek*, will likely need to be replicated rather than simulated, as there's no known way (yet) to simulate the feeling of being wet without the subject actually being wet in reality. While the actual execution of a holodeck might look extraordinarily different than a graph-paper-like room powered by holoprojectors, replicators, force fields, and tractor beams, the fact remains that this technology is being rapidly developed and brought toward fruition. By the mid-twenty-first century, perhaps all of us will have the opportunity to choose our own holovacations, and holoprogrammer will become a bona fide profession!

THE SHIP'S COMPUTER

"Auto-destruct cancelled"—a sentence that offers the crew a huge sigh of relief when it's your ship's computer that utters it. The ability to talk to, be understood by, and be obeyed by an artificially intelligent machine is quite a dream, especially when you consider that the machine, wired to the starship, is literally the only thing separating you from the lethal abyss of deep space. Simultaneously, a ship's computer is your interface to virtually unlimited computational power, the entire storehouse of your civilization's knowledge about the universe, and the controls over every one of the ship's systems, from shields to engines to life support. But the most remarkable advance of a ship's computer is your ability to speak to it in your native language, as naturally as you'd speak anything that was on your mind, and have the ship engage its full computational abilities toward whatever end you command it, so long as you have the proper authorization.

The ship's computer is integral to doing just about anything onboard a starship.

A starship's main computer is an incredibly powerful but incredibly resource-intensive behemoth. The *Enterprise* NX-01 had a core three decks high that was the most powerful computer ever created by the mid-twenty-second century. A hundred years later, duotronic computers had taken over, and a computer access room was a part of every *Constitution*-class ship, to provide easy access to the large computer core. By the twenty-fourth century,

The greatest advances in computational power have come from continued miniaturization, not from scaling near-future technologies up, as envisioned by *Star Trek: The Next Generation*.

duotronic computers were replaced with isolinear ones, with smaller ships like *Voyager* containing two redundant, independent cores and larger ones like the *Galaxy*-class *Enterprise*-D containing three, all interconnected by an optical data network. The mechanisms, scale, and computational capabilities these systems possessed were unheard of at the time they appeared in their respective *Star Trek* series.

The fields of machine learning and natural language processing are well on their way to emerging from their infancy and into our daily lives. Already, smartphones running iOS and Android have native voice-recognition applications capable of identifying voice inputs in dozens of languages with vocabularies of tens of thousands of words each. Those voice inputs can be applied to a number of different ends, from scouring the Internet for related information to translation into another language (see page 99) to computing the ideal walking, driving, or public transit path to an arbitrary destination. Just as you'd expect from *Star Trek*–worthy technology, it can give you the appropriate outputs to your inquiry in a combination of formats: text, graphics, or audio, all at once in some cases. And with the emergence of self-driving cars, soon you won't even need an ensign at the navigation consoles to lay in a course to your destination—a simple voice command in your vehicle will automatically transport you there along the optimal path.

But the ships' computers in the *Star Trek* universe go far beyond the capabilities of a smartphone's personal aide. While systems like Siri, Google Now, and Cortana only provide information they can get by querying a database (like the Internet) in a particular, straightforward fashion, *Star Trek*'s computer displays a complex understanding of a query's context. It understands when to truncate information for accessibility, when to be verbose to provide additional context, and when to cease its outputs—even midsentence—to accept a new (or refined) query. It even appears to have a memory of which previous queries have failed, along with how they failed, which informs how it responds to future commands in a user-specific manner. The rudiments of

What took a whole room to compute at the time *Star Trek* premiered, as pictured here, can now be accomplished in a handheld device.

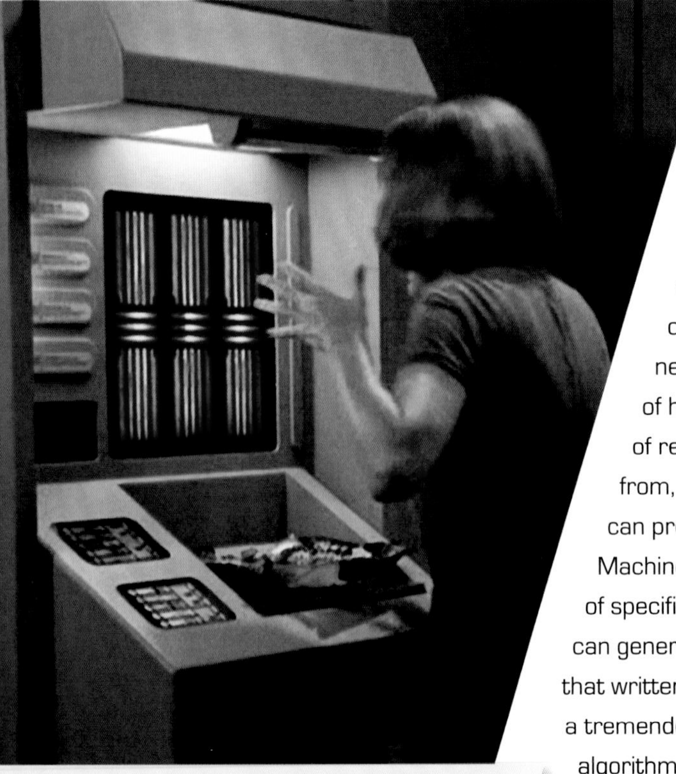

A computer that programs itself to obtain all the necessary, relevant information desired by the user might be considered quite intelligent. But if it fails to deliver a message like "Warning: plate is hot" before the human in question touches the hot plate, it isn't performing quite as desired.

this technology, in the form of personalized searches or "smart search" technology, are already available today but not widely incorporated into digital assistant programs.

The major development that will be necessary to have a computer truly capable of what *Star Trek* computers are is the ability for a computer system to write substantial amounts of its own *unique* computer code. At present, even artificial intelligence systems need to be programmed with overwhelming amounts of human-written code. Although there are databases of results and responses that a machine can "learn" from, there is only a prescribed set of parameters that can presently be adjusted, set by the human programmer. Machine-learning algorithms can generate code to some set of specifications, and on occasion a machine-learning system can generate code that outperforms or is more efficient than that written by human experts. But scaling has proven to be a tremendous problem; right now computers can only write algorithms to solve very specific classes of problems they've been programmed to address. Even then, they can only write maybe a maximum fifty to one hundred unique instructions in a single program.

Even with the vast computational power at the disposal of the world's most powerful research laboratories today, computers are not yet capable of writing large amounts of novel code on their own. To achieve levels of sophistication beyond its initial programming, a computer must be able to create its own subroutines. To perform any computations or scans outside of the ones it was programmed to

No matter how sophisticated a computer system is envisioned to be, it can never be expected to do everything, nor can it ever fully replace what human beings can do.

accomplish initially, it must be able to process an inquiry, break it down into a series of algorithmic steps, encode and perform the computations that must take place, and then relate the results, in natural language and in a timely fashion, to the user. A failing on any of these fronts will result in the same apparent result from the perspective of the human interacting with the computer: an incompetent machine.

Of course, even in the twenty-fourth century, there are limits to what a computer can accomplish without further programming. Direct interfaces with a human brain—a computer capable of responding to mere thoughts—are only in the very preliminary stages here on Earth today, yet were foreseen to still require instructions to the ship's computer more than three hundred years in the future. At some level, no matter how artificially intelligent a computer gets, there may still be some programs that can only be written by an external programmer.

In addition to playing Christine Chapel and Lwaxana Troi, Majel Barrett—actor, producer, and wife of Gene Roddenberry—is the iconic voice of the shipboard computer in *Star Trek*'s series and films.

Thanks to the advances in computational systems—both in programming and in raw power—we are much closer to a computer as envisioned by *Star Trek* than even Gene Roddenberry might have predicted we'd be by this point. Computers can receive and process voice-activated commands, compute the results with a combination of preprogrammed code and a limited amount of their own self-written code, scouring the entire landscape of digitized human information, and relate those results back to users in their own native language. What's available today might not quite live up to the *Star Trek* vision in every aspect just yet, but one prediction about the future has remained true ever since the first home computer came online: users will forever be frustrated with the limitations inherent to the particular system with which they interact, no matter how powerful it comes to be!

PADDs

When you're traveling through the galaxy in a starship, you're never out of reach of the ship's computer or very far from a tied-in console. But sometimes, the work you're doing is best done using a handheld device. Maybe you're on the move and need to take your work from one deck or station to another, or you have to deliver an important set of computerized information in person to your superior officer. Even in the far future, computing needs to be personal in a way that was unfathomable when *Star Trek* first aired. The smallest computers in the 1960s were the size of entire rooms, and so the idea that one could simply carry around a device that was tied into the powerful shipboard computer in handheld form, yet that could be worked on with a stylus just as easily as one would work at a console, was somewhere in between futuristic and magic. By time the first episode of *The Next*

What appeared to be detailed layouts or technical diagrams on a computer display were actually static prints, lit or unlit, behind a sheet of plexiglass. Yet the technology behind these "Okudagrams" actually came to fruition just a few years later.

While the idea of a computationally powerful notebook-sized device was futuristic in the 1960s, it was the vision of *The Next Generation* that largely inspired the tablet computers that are so widespread today.

Generation aired in the late 1980s, the original series' "electronic clipboard" had evolved into a full-fledged interactive computer: the Portable Auxiliary Display Device, or PADD.

PADDs varied in size, shape, and specialization functions, and were widespread among spacefaring civilizations. Federation, Klingon, Cardassian, and Ferengi versions all had their own unique configurations, and were perhaps the first handheld touchscreen devices ever envisioned. Some featured buttons to provide tactile feedback and a variety of controls; others had larger screens and styluses for more detailed manipulation and could be used for anything from log entry and blueprint display to image reconstruction, video communication, piloting, and much more. In short, they were mobile, personally accessible, massively powerful computing devices that were as common aboard a starship as an officer's uniform.

Despite how futuristic they seemed when they were first envisioned and placed on screen with the debut of *Star Trek: The Next Generation*, the technology envisioned in PADDs came to fruition far more quickly than even the most optimistic experts predicted. Clearly, computers of the era were getting faster and smaller, with not only text interfaces but much-improved graphical capabilities. Still, an interactive, powerful, multifunctional computer with touchscreen capabilities seemed incredibly far off at a time where the personal computer was mostly a large desktop tower with a CRT monitor used for word processing. Yet just a few years after *The Next Generation* premiered, touchscreen PDAs began to become widespread, followed by the rise of mobile phones and, led in large part by Apple in the late 2000s and early 2010s, tablets and smartphones. In only a few decades, the PADD had become not only real but ubiquitous, with more than a billion touchscreen "smart" devices active worldwide, and its name is is even conjured up by that of the most successful tablet of all: the iPad.

The accumulated knowledge and information discovered by all of humanity is now nearly all accessible online, with practically all mobile devices worldwide able to freely access this data. Even though *Star Trek* didn't envision such a system, and PADDs and electronic clipboards instead relied on the vast encyclopedic knowledge of the memory banks on the main ship's computer, the series understood the importance of handheld devices tying into a larger database. They could be used for text communication; they had autocomplete functionalities for words and sentences; they could be used for computationally intensive tasks just like a ship's console could. Of course, today's smartphones and tablets can do all these things and more. The only things they can't do that a PADD

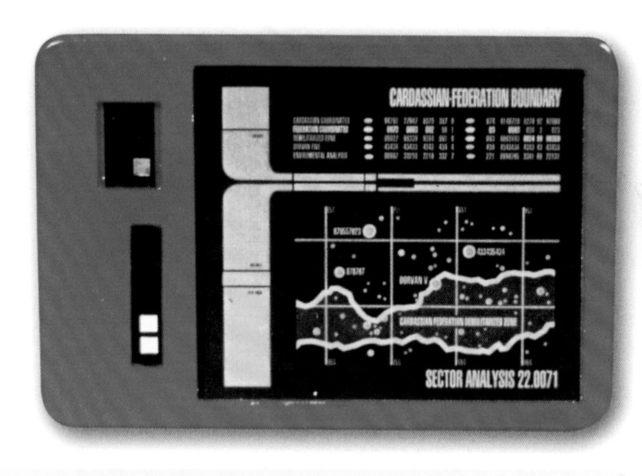

By the time 2010 came along, tablet computers had evolved to the point where they had brought to life pretty much everything imaginable in *Star Trek*.

could is pilot a starship or initiate a transport—and that's only because the average tablet user doesn't have starship to pilot.

The major breakthrough that allowed this technology to become a reality was a shift from text-based computing, through command-line interfaces like DOS or UNIX, to an intuitive graphical user interface, pioneered by the Apple Macintosh in the mid-1980s. When the iPhone and the iPad came out two decades later, they represented another leap of even greater magnitude. Their interfaces were much more intuitive to use for many people than other tablet-like devices at the time, and even technophobes were quick to embrace how quick and easy they were to use. In many ways, that was the real goal of this technology from *Star Trek*'s perspective: not that you'd have such a powerful device in your hand, but that you'd have such a beautiful, useful device that became crucial to your life once you'd begun to use one. The aim was instant information and technological gratification anywhere you went. Although there are always more improvements, refinements, and even revolutions to come in the future, the technology of the PADD is here to stay.

The ability to access the full suite of information available to humanity at your fingertips has been a dream of people since ancient times, long thought to have been out of reach. Today's pocket-sized smartphones and notebook-sized tablets allow us instant communications in a variety of formats or access to images, videos, and a suite of information that would make any twentieth-century librarian's jaw drop to the floor. Beyond that, they connect us to the entire world and even to astronauts in space, just as communications aboard a starship connected not only the officers and the crew, but also planets and species across the galaxy, and even artificial life-forms and holograms.

The real-life applications available on a tablet or smartphone, from games to browsers to productivity apps and everything in between, already far exceed anything *Star Trek* showed us on a PADD. Yet it was arguably the light-up user interfaces displayed on the earliest PADDs and consoles—officially called the Library Computer Access/Retrieval System (LCARS) but affectionately known by many *Trek* fans as Okudagrams after their designer, Michael Okuda—that provided the inspiration for the easy-to-use design of tablets and smartphones we know today.

Advances in LED displays and their successors (the blue LED, which was awarded the Nobel Prize in Physics in 2014, was only first invented in 1994) coupled with increased computing power and miniaturized components have paved the way for smartphones and tablets to surpass the incredible dreams of *Star Trek*. This technology is such an integral part of our daily lives that it will take another brilliant revolution—perhaps in the form of a computerized, thought-controlled ocular implant—before they fall out of favor.

In an advanced enough android, an entire human brain could be theoretically transferred into the automaton's body. If this were possible, the question of whether this is actually still you or not becomes a fascinating and terrifying one.

ANDROIDS

An android, "an automaton made to resemble a human being," according to *The Next Generation*'s Data himself, is a completely mechanical humanoid robot. As envisioned by *Star Trek*, these machines would appear as actual humans from the outside. They would move like us, talk like us, and—if their programming allowed it—feel, think, deceive, and even love like us. But these wouldn't be cybernetic organisms with living tissue, or modified humans or humanoid creatures in any way. These are purely artificial creations, albeit with their own thoughts, minds, and arguably, sentience and consciousness. With completely artificial, computerized bodies and minds, they could conceivably outclass anything that the natural evolutionary process has brought about by any metric.

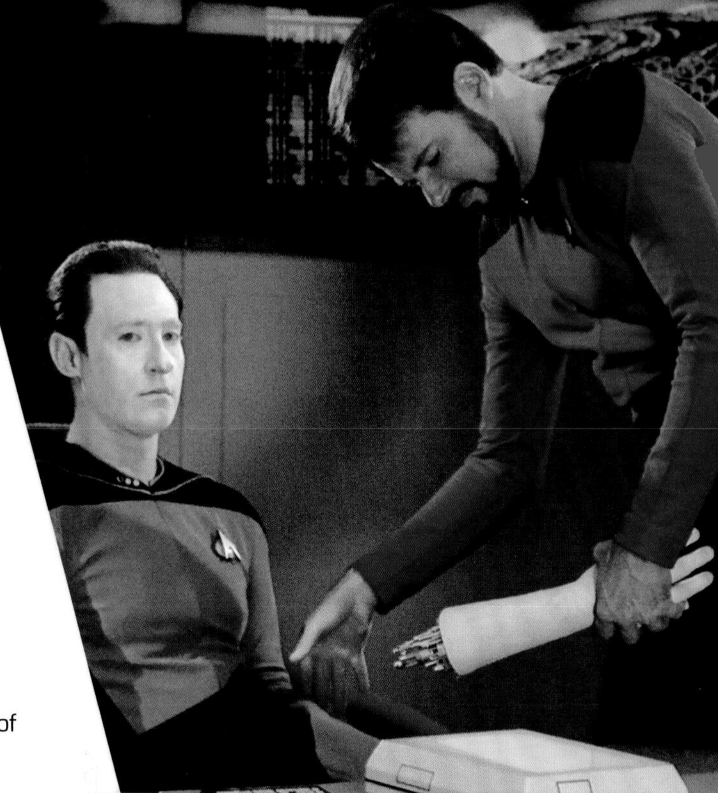

The line between an artificial life-form and a mere machine is quite unclear, and remains without a definitive answer today.

While humanoid androids were referred to in the twenty-second century, they were never seen. The crew of the original *Enterprise* and *Enterprise*-A, however, encountered a great many androids originating from various cultures, including those created by Flint, Mudd, Sargon, and those from the planet Exo III. But no advanced androids originating from the Federation came into existence until Dr. Noonien Soong made his breakthroughs in the twenty-fourth century. Culminating in the creation of Data and Lore—androids that had superhuman physical and computational abilities but were missing a few important aspects that would make them fully humanlike—the most advanced androids appeared to have their own autonomous minds. Whether they were truly alive and sentient or not, the way we consider ourselves to be, was a question that *Star Trek* was happy to repeatedly address, but never definitively resolved.

To create an android as envisioned by *Star Trek* requires two equally important but independent components: the body and the mind. An android body without any programming inside it is simply a blank; a computational mind without a body is simply a computer program. Yet if you put them together, the result is the most humanlike creation outside of an actual human being. In 1964, animatronic creations—robotic humanoid characters with lifelike, preprogrammed motions that moved in time with human voice recordings—debuted at the World's Fair in the form of Abraham Lincoln. For the first time, humanity had a glimpse of what an android could truly be, despite the copious workarounds of the actual technical limitations at the time.

For a long time, the android face and body were stuck in the deepest pits of what has been termed the *uncanny valley*, where these mechanical creations looked just enough like a human to cause a visceral reaction of repulsion in those who viewed it. But beginning in the mid-2000s, advances in robotics coming from Japan and South Korea led to the first androids whose facial expressions changed appropriately when they spoke, moved, and reacted to their environment. A variety of joints, pivot points, and other intricacies allowed these primitive androids to display humanlike characteristics and gestures, even as they interacted with a human participant. While none

The Japanese android Actroid-DER, developed by Kokoro and pictured here in 2015, was a rudimentary human replacement for various customer service needs in four languages, with nearly fifty independent pivot points in its face.

of the androids designed to date would ever be mistaken for a human upon any sort of close inspection, the progress made on this front in recent years has been impressive.

While lifelike humanoid bodies may be close to a reality, it's imperative that these androids not be mere husks that are pretty to look at but nonfunctional. But as computational power and programming continue to improve, they can be seen to accomplish tasks in a superior fashion to any human. Human chess players have not won even a single game against the top chess programs under match conditions since 2004, while Google's AlphaGo program defeated one of the top human go players in 2016. But "body intelligence" is fast improving in androids as well. The annual RoboCup soccer match has seen tremendous progress in the arena of soccer-playing robots, and it's estimated that by the 2040s, the robots will be competitive with top humans. As artificial intelligence, machine learning, robotics, computing power, and programming nuances all continue to advance, it may not be long before androids are outperforming humans at practically all tasks we presently engage in.

The humanlike "brain" of an android envisioned by *Star Trek* is the most elusive component of this technology, but it may yet be both physically and technically possible. If so, perhaps an advanced enough android could even reproduce.

Even with these developments, there's still a big difference between a humanoid robot that moves like a human and outperforms humans at many tasks, and what we envision as a true *Star Trek*–style android. The biggest difference? A unique personality, a consciousness, a sense of self-awareness. Although that isn't here today, perhaps it isn't as technologically far away as we might imagine. Just as humans are predisposed toward any number of traits but can choose how they act, perhaps the proper algorithm in an android that perceives, processes, and reacts to external stimuli can wind up creating its own personality. Perhaps programming in options that allow an android's experiences to inform identity and advocate for its own self-preservation would impart self-awareness. And just maybe, machine learning and artificial intelligence in a neural net setup can someday replicate what we understand as consciousness. There isn't a clear path to these end goals as of today, but the field of cognitive computing is working on it.

The holy grail of android technology is reproducing what we consider the finest abilities of the human brain:

- to gather ambiguous information across multiple senses
- to create abstract mental categories such as time, space, and objects
- to form interrelationships and associations between seemingly unrelated stimuli
- to synthesize it all together to formulate a novel idea and a course of action

Currently, the limitations are that a computer program can only work within a set of bounds imposed by the programmer; the maximum amount of learning possible is limited. But if we can supersede those limits—if we can create a truly artificial intelligent mind—we will not only have realized the *Star Trek* dream of creating a sentient android, we will, from a particular point of view, have arguably created a new form of life.

Unlike most other technologies envisioned by *Star Trek*, however, this one brings up a slew of not just technical but ethical questions. If it's possible to create an android that truly reaches the levels of intelligence, self-awareness, and consciousness that *Star Trek* envisioned, will we have created life? What rights will we be required to grant such androids? Would it be ethical—if possible—to duplicate an android, or even replicate or download the mind of a human being into an android body? And if so, at what point do we cross the line from nonlife into life, from property to autonomy, from something with no rights to something with inalienable rights? As Captain Picard said, when advocating for Data's right to choose a path for his own life, "It could significantly redefine the boundaries of personal liberty and freedom, expanding them for some, savagely curtailing them for others. Are you prepared to condemn him and all who come after him to servitude and slavery?" As our androids get more and more advanced, or become more humanlike, at some point we're going to have to draw that line for ourselves.

The culmination of Noonien Soong's life work are his most advanced androids, Data and Lore. These two had vastly different personalities, moralities, motives, and lives, yet were created out of the same parts.

CIVILIAN

TECHNOLOGY

Above all else—above the warp drive, the photon torpedoes, the transporter, the medical marvels, and the technological revolutions—*Star Trek* is about envisioning a better future for the benefit of all of humanity. A society devoid of hunger, poverty, and preventable disease was just the beginning, as *Star Trek* brought us the idea of gender and racial equality for all, and even extended it to alien species, the final frontier in what an "other" might look like.

In the *Star Trek* universe, the quality of life hasn't risen to a remarkable level just for those on a starship, but for everyone across the United Federation of Planets. Security, freedom, leisure time, and entertainment are plentiful, as technology has freed humanity from the burden of needing to be continuously productive to earn a living. The acquisition of material goods and other assets of value is no longer synonymous with phrases like "the pursuit of happiness," as the great technological benefits of the far future are accessible to all. Even something as simple as walking through a door is in some ways a futuristic experience in *Star Trek*—at least, it was in 1966.

While it is perhaps an absurd idea we could reach a point where no manufactured goods or fabricated parts would require a skilled human to help create them, *Star Trek*'s replicators foresaw a future where anything could be created with a simple command. That day is not far off, with 3D printing for inorganic materials having already become a reality, and with 3D-printable food in a moderate stage of development. Even the ability to drink an intoxicating beverage as much as you like and not suffer any ill effects is something that we won't have to wait for until the twenty-fourth century. There's never been a better time to be a civilian, and with the emerging *Star Trek* technologies coming to fruition, it promises only to get better from here.

Captain Picard was well known for his replicator order of Earl Grey tea.

REPLICATORS

"Tea, Earl Grey, hot," demands Captain Picard, with as little wonder as we might express at turning on a toaster. In a matter of perhaps two seconds, a teacup appears out of thin air, filled with tea brewed to perfection. Only a voice command away, the replicator replaces our need for manufacturing, for the ownership of goods, for food and drink; all of it can be delivered with a simple order, out of nothing but thin air. This visionary technology relies on the simple, long-known principle that all forms of matter are made of the same fundamental building blocks—the same protons, neutrons, and electrons configured into different atomic combinations—that only differ in how they're bound together.

"Tea, Earl Grey, hot."

The twenty-fourth century, in the *Star Trek* universe, saw a superior replacement for the original series' food synthesizer: the replicator. By creating atoms, then molecules, then more complicated structures and configurations (from proteins and cells to full cuts of meat) out of their subatomic components, anything imaginable could be created so long as a sufficient model existed. The only constraints seemed to be the amount of available energy required to put it all together, which is why the crew of the lost *Starship Voyager* was put on replicator rations, denying Harry Kim the simple pleasure of playing the clarinet until he saved enough rations to replicate one. In perhaps the boldest replicator use of all, a hostile species kidnaps Captain Picard and replaces him with a replicated doppelganger, testing the crew of the *Enterprise*-D.

Harry Kim's clarinet was one of his most cherished possessions, left behind in the Alpha Quadrant despite his mother's pleas to Captain Janeway. He was only able to replicate one after saving up enough replicator rations.

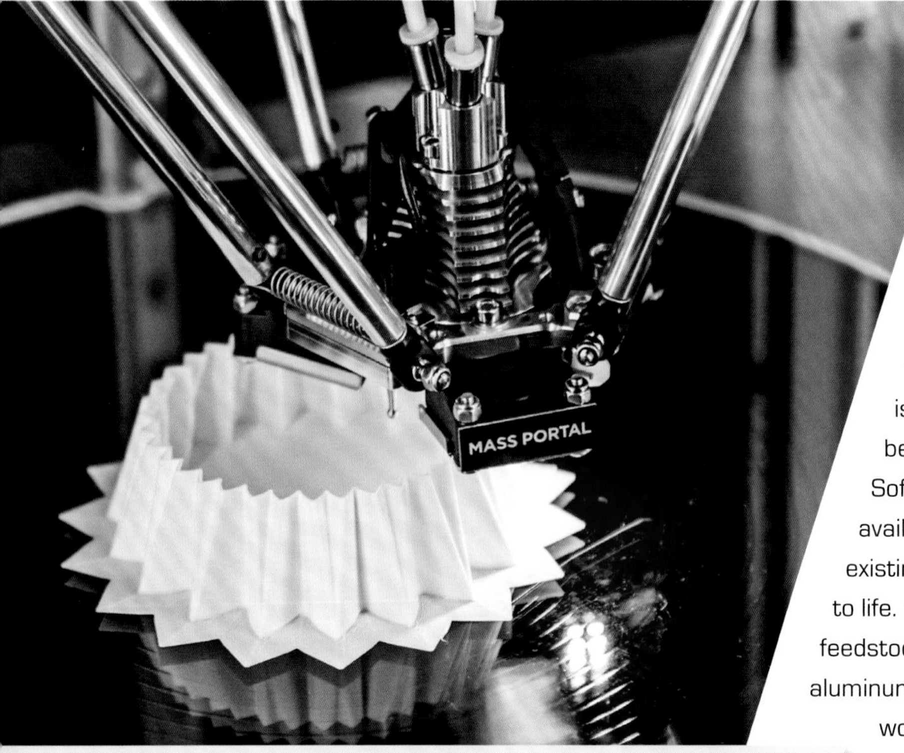

3D printing technology is quickly becoming better and cheaper. This printer was demonstrated at the 3D Expo 2016 in Moscow, Russia.

While we might be a long way from replicating perfectly cooked food, and a living creature is far more distant still, the ability to replicate teacups, clarinets, and all sorts of inanimate objects is already here. 3D printing, in which any modeled object can be printed out of whatever feedstock is inserted into the printer, is not only commercially available but becoming increasingly more affordable. Software modeling and 3D scanners are available as well, allowing any physically existing or designable object to be brought to life. Originally only able to use plastic feedstock, 3D printers can now print foam, aluminum, carbon, titanium, steel, and even wood structures. With the appropriate models and printers, they can replicate everything from art to tools to clothing to entire houses.

By printing out any continuous object in thin, additive layers, one of these devices can build up a structure in any configuration at all. Different 3D printers with different feedstocks will print to different levels of precision; the thicker options offer imperfections that are nearly millimeter-scale, clearly visible to the naked eye, while more precise combinations can create objects with layers down to a thinness of around 20 microns, or one-fifth the width of a human hair. The flexibility of 3D printers to create any intricate, continuous pattern allows practically any design to be input, including those for objects whose physical construction from raw materials would otherwise be impossible. Ornate, fashionable accessories are just as possible as puzzles, sneakers, casts, miniaturized vehicles, gears, sculptures, or seemingly Escher-like impossibilities brought to life. If you can design it, it can be replicated—people have even used this technology to create a working lawnmower! 3D printers also have extraordinary medical applications, ranging from prosthetic limbs and attachments to custom joint and bone replacement materials. And humans aren't the only ones who have benefited: 3D printing has been used to replicate replacement shells for sick or injured tortoises.

This technology is, at present, limited by the materials that have been developed to print with. However, advances in quantum dots, nanoparticles, and stereolithography have the potential to create new epoxies and the ability to 3D print using composite materials. This would allow a greater diversity of materials to be used (and used together), and enable the creation of finer layers and coatings, perhaps down to submicron precision. The possibility of making 3D-printed graphene structures is close on the horizon, because we're making these sheets smaller and thinner as the years tick by. Most projections indicate that printing a layer the thickness of a single atom may be possible during our lifetimes.

But right now, we can't envision making water out of plastic, nor can we make the flavor compounds created by steeping tea leaves out of anything other than tea leaves themselves. While we're well on our way to figuring out how to create feedstocks out of any conceivable inorganic material and a few organic

For Starfleet crew, onboard meals came from the self-serve replimat. Here in the twenty-first century, 3D-printed foods are already being designed for astronauts aboard the International Space Station.

ones (notably wood), we can only make replicated food out of feedstock that is itself made of foodstuffs. While dehydrated food powders can be reconstituted and used as feedstock, as can liquid or semiliquid foodstuffs such as chocolate, we have not yet cracked the nut of turning nonfood into food. This is a technical challenge that may someday be overcome, but would require the ability to manipulate matter at a much finer, more granular level than our current precision allows.

3D printing a model of a spine is easily accomplished by 3D printers of today, but printing a vertebra good enough to replace a damaged one in a human body is not yet routine. This spine model with a 3D-printed artificial axis was the first to be implanted in a patient, a surgery carried out by Chinese doctor Liu Zhongjun.

The food synthesizers imagined by the original *Star Trek* provided a method for humans to survive for long periods in space without the need to worry about agricultural shortfalls or devote tremendous resources to simply keeping basic nutritional needs met. *The Next Generation* went a step further, dreaming up a future in which any object—inorganic or organic—could be replicated, given the proper inputs. We've now demonstrated that food synthesis is possible, with 3D food printers presently an emerging technology. Inanimate objects, however, have already reached a level that would satisfy the discerning eyes of Picard, Crusher, or Riker, if not quite the more sophisticated senses of Data or La Forge. If we can overcome the barrier of turning nonfoods into edible, nutrition-rich delectables, we can safely say that this technology will have become reality, and done so centuries ahead of *Star Trek*'s predictions. After all, if the *Starship Voyager* were equipped with today's technology and the proper feedstocks, Harry Kim wouldn't have had to wait—or sacrifice his replicator rations and suffer Neelix's questionable cooking—for the joys of his clarinet.

NASA launched the first 3D printer to the International Space Station in 2014. Built by California company Made in Space, it is designed to work in the ISS's zero-gravity environment.

Only the most refined connoisseurs of alcoholic beverages had palates that could discern the difference between alcohol and synthehol.

SYNTHEHOL

"I'm a joke on Ferenginar. Starfleet's favorite bartender, the Synthehol King!" Yet as much as Quark might have lamented his lack of prestige for being a slinger of synthale, there is no shortage of intoxicating beverages across many civilizations and time periods in *Star Trek*'s version of our galaxy. From Captain Archer's fondness for Andorian ale to Kirk's and McCoy's immediate onset of drunkenness after consuming Romulan ale, the Klingon staples of bloodwine and *chech'tluth* introduced by Worf, Ten-Forward classics like the Samarian Sunset, and Gul Dukat's extortion of Quark with payment to be made in cases of Cardassian ale, the effects of alcohol and alcohol-like beverages play a major role in *Star Trek*. While many rituals and professional events—both celebrations and solemn moments—call for its consumption, the vast majority of these drinks were consumed when the crew was off-duty, in a more comfortable and less professional environment.

Yet the enjoyable sensations these drinks provide, such as intoxication, improved confidence, enhanced wisdom, that "warm" feeling, and some measure of perceived invulnerability, come along with downsides that are not so easily dismissed. Drunken crew members would find themselves severely impaired if called to duty in an emergency; the risks of addiction and toxicity are still severe, even with improved medical care; the sheer discomfort of drinking too much is something no one enjoys experiencing; and the hangovers can stay with you all throughout the next day. As Montgomery Scott once noted, "Never get drunk unless you're willing to pay for it the next day." But while alcohol's negative effects were a necessary evil for anyone who wanted to partake of its delights for the crews of Archer and Kirk, the twenty-fourth century brought an alternative: synthehol.

Dreamed up by Gene Roddenberry and credited to the Ferengi for its invention, this synthetic version of alcohol promises to deliver all the pleasant effects of alcohol, yet none of the drawbacks. Imagine that one moment, you're enjoying yourself with your friends, laughing and carrying on in your tipsy state, while the next instant, the red alert alarm goes off. Instantaneously, as the rush of adrenaline flows through your bloodstream, the tipsiness disappears, and you're back to a stone-cold-sober state. No hangover, no upset stomach, no loss of equilibrium, not even a hint of dizziness or blurry vision.

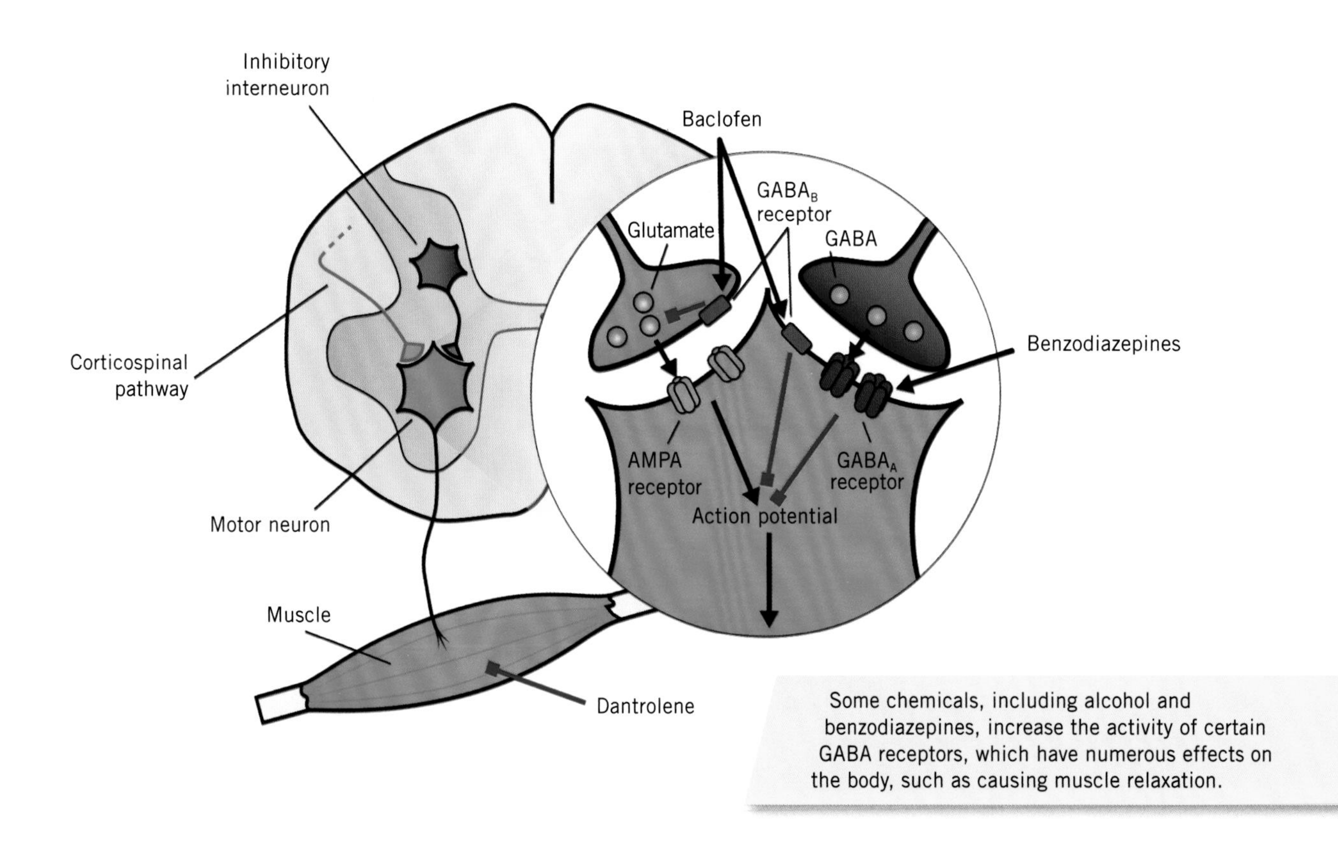

Some chemicals, including alcohol and benzodiazepines, increase the activity of certain GABA receptors, which have numerous effects on the body, such as causing muscle relaxation.

With no danger of addiction or toxicity, it sounds like one of the greatest public health breakthroughs a modern society could ask for. While the mechanism proposed by *Star Trek*—that almost any humanoid could break down the synthehol almost instantly, even at will in some cases—may not be feasible, perhaps the development of a nontoxic but still enjoyable substitute for alcohol isn't destined to remain fiction forever.

There are entire classes of chemicals that give a feeling of relaxation, well-being, and antianxiety when ingested by humans, including alcohol as well as benzodiazepines like Valium, Xanax, and Klonopin. There's a good reason for this, as the neurochemistry of all four drugs has something in common:

they're all agonists (or enhancers) of gamma-aminobutyric acid (GABA), a neurotransmitter that depresses the central nervous system. It has a sedative effect, causes sleepiness, and gives a more relaxed feeling. There are other effects that GABA has, however, that one may *not* desire as much, including memory loss and nausea, depending on which neuroreceptors the GABA molecule binds to. The ultimate goal of a safe but effective alcohol substitute would be to minimize the negative effects that GABA and acetaldehyde (the broken-down form of metabolized alcohol that causes hangovers) have while still keeping that relaxed, buzzed feeling. While alcohol and the benzodiazepines are full GABA receptor agonists, meaning they enhance the effects of GABA across the board in the brain, it should be possible to develop a chemical that acts as a *partial* GABA receptor agonist instead.

In general for ligand-gated systems—the ones activated by molecular bindings—known partial agonists have a weaker effect on the channels they bind to. In theory, when it comes to the channels activated by GABA, it should be possible to produce an agonist that only activates the pleasurable receptors (for pleasure, relaxation, and that "buzzed" feeling) while leaving the undesirable ones (for nausea and memory loss) inactive. And in the case of all benzodiazepines, instead of producing the hangover-inducing acetaldehyde, a simple dose of a readily-available drug—flumazenil—serves as a near-instantaneous antidote, restoring the person who takes it to full sobriety in very short order. (That said, flumazenil itself results in dangerous side-effects to many patients.) Modern advances in psychopharmacology have made huge strides toward finding a safer, better alternative to alcohol, with one leading candidate being the chemical bretazenil. Invented in 1988, it's a member of the benzodiazepine family that acts as a partial agonist, is less resistant to drug tolerance, has few withdrawal symptoms,

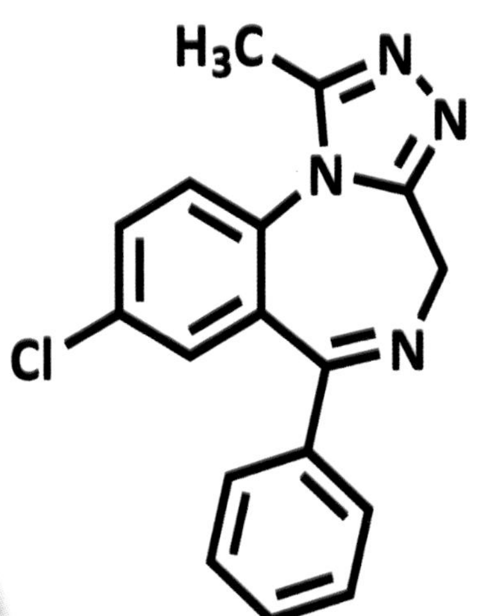

The molecular structure of the generic family of benzodiazepine molecules, in which a variety of slight to moderate modifications can lead to different partial or full GABA receptor activations.

and, most compellingly, binds to specific GABA subunit receptors (α_4 and α_6) that other benzodiazepines don't! While bretazenil, originally designed as an anti-anxiety medication, offers the positive relaxation and sociability effects that alcohol does, many of the negative effects, including aggression, amnesia, nausea, loss of coordination, liver disease, and brain damage are simply not associated with it. By developing a drug that binds to the desired GABA subunits but not the undesired ones, where the full suite of effects could be turned off in under an hour, it's conceivable that a real-life version of synthehol may be just a drug trial away.

While it might seem like a dystopian future to Scotty, who found a twenty-fourth century filled with "synthetic scotch" and "synthetic commanders" a wee bit hard to adapt to, the development of a safe, enjoyable, and easily reversible alcohol substitute would be a tremendous boon to health and safety worldwide. Successful real-world development is more dependent on government regulations than any scientific constraints at this point, as it seems to merely be a matter of finding and legalizing the right chemical to do the job. Just as alcohol gives many of us the courage and confidence to sing, dance, or approach a potential love interest, synthehol does the same for inhabitants of the future without any of the negative health repercussions. With the development of an easily broken-down alcohol substitute, not only would we be able to drive ourselves home after a night at the bar with no danger, we wouldn't have to worry about alcoholism, cirrhosis, alcohol poisoning, hangovers, or many of the other ill effects alcohol has caused throughout our history. The only thing it wouldn't cure is the shame we might feel after sobering up!

In *The Next Generation*, Guinan was the purveyor of synthehol at the Ten-Forward bar.

A door that can infer your intentions from your motions, and know when and whether to open and close or not, was one of the most novel domestic technologies envisioned by *Star Trek*.

AUTOMATIC SLIDING DOORS

With a *swoosh*, the automatic sliding doors part ways to let you through. But only, somehow, if you intend to go through them. Walk faster toward them, and they open up more rapidly; run toward them and they'll be open when you're there. Once you pass through, they'll close behind you, unless you turn around to say something to those in the room you just departed, in which case they'll stay open long enough for your to finish your conversation. (Or, if such is your wont, just to get your final zinger in.) Yet if you walk in their vicinity but don't want to leave the room, even if you walk right up against the door itself, they remain closed. A door that will open when you need it to, at the speed you need it to, and remain closed (or to close behind you) only if you want it to might sound like some type of psychic piece of magic or witchcraft, but in the world of *Star Trek*, if you could imagine a technology, Gene Roddenberry and his successors had a way of bringing it to life.

The ability for inanimate objects to be sensitive to your context is a technology that predictive modeling is a necessary prerequisite for.

This may seem like one of the least impressive pieces of technological innovation that *Star Trek* has given to humanity, since not only are sliding doors common everywhere these days—from airports to hospitals to elevators to supermarkets—they're not even all that good at what they do. Yet what we have today in the twenty-first century is a direct result of a technology envisioned by *Star Trek* having become so influential and ubiquitous that many of us alive today have no memory of a world where sliding doors didn't exist! The convenience

The frames and tracks for sliding doors have been found in the ruins at Pompeii.

of being able to enter and leave a space hands-free, is something that could easily become a convenience we take for granted. Aboard a Federation starship, automatic doors enabled crew members to focus on all the other tasks inherent to their lives and their jobs at all times, without having to pause to clear a path for themselves. Instead, the door senses what you're doing and gives you exactly what you want. This sensitivity to your context—staying open for longer when you linger and closing rapidly when you're escaping from something—makes *Star Trek*'s doors unique.

The dream of doors like this didn't begin with *Star Trek*, but rather a decade earlier, as the brainchild of Dee Horton and Lew Hewitt in 1954. (Manual sliding doors, of course, are nothing new—they have been in use since at least the first century in the Roman Empire.) The original idea behind automatic sliding doors was that you could put a weight onto a sensor that would make electrical contact, completing a circuit, and cause a door to open as a result. The linear motion—either vertically or horizontally—was an innovation to deal with the windy conditions Horton and Hewitt faced in Corpus Christi, Texas. Six years later, the Horton Automatics company was founded, and soon sliding doors became widespread in public spaces such as shopping malls. Later on, piezoelectric foot sensors brought about greater sensitivity, motion-sensor technology enabled a door to only open if an object moved, and timed circuits enabled automatic closures after the proper, preprogrammed amount of time had passed.

Still, nearly two decades into the twenty-first century, most automatic doors in operation operate on two crude infrared sensors in the proximity of the door, which can be triggered by any substantially large object, animate or inanimate. The sensor in front of the door causes the door to open while the sensor at the door causes it to remain open, otherwise a timed circuit causes the door to close, with all openings and closings occurring at the same speed. This is a far cry from the true dream of a *Star Trek*-style door, which appears to come complete with the intelligence aspect: for it to somehow read your intentions and open or close accordingly, at a speed that's appropriate for your activity level. In 2014, a team of researchers at Japan's University of Electro-Communications and Hokuyo Automatic Company invented a smarter automatic door that takes advantage of 3D motion-sensor technology, using laser scanners, coupled with computer vision algorithms. Similar to technology found in introductory

Sliding doors can either remain open indefinitely, close when someone leaves the sensor area, or operate in a timed fashion, depending on how they're programmed.

Automatic sliding doors on an elevator are certainly a familiar sight.

physics (mechanics) laboratories or, more familiarly, in sensors like that of Microsoft's Kinect for Xbox consoles, it can detect numerous individuals at once, their instantaneous positions, and their velocities and accelerations (including directions)—meaning that it can then be programmed to open and close appropriately, including at the optimal speeds and even by the optimal width. This new technology enables the door to be kept closed at virtually all times, except when a human intends to pass through it, including in all practical weather and lighting conditions.

Right now, the only thing preventing widespread implementation of this smarter door technology is cost: the additional sophistication in sensors and motors comes at a cost of approximately $1,000 extra per door. Although this promises long-term savings in terms of lowering the heating and cooling costs of a given space, as well as providing the privacy of having more securely and consistently separated rooms, the technology has yet to catch on in residential or commercial settings. The addition of uniquely identifiable proximity devices, such as smartphones, beacons, or other Bluetooth-equipped gadgets, could further enable only authorized personnel to activate such an automated sliding door, while securely keeping others restricted to their side of the door. Future developments may apply facial recognition technology to door triggers, enabling them to become a routine part of tomorrow's "smart homes."

With the exception of those looking to "fool" the door by running toward it and then stopping short, causing it to open in unintentional situations, the sliding door technology envisioned by *Star Trek* is already a reality here on Earth today. The version that we commonly encounter in our daily lives is many decades out of date, as what we're capable of lives up to the version envisioned by *Star Trek* almost completely. In fact, a large number of developments promise to have this technology soon outstrip anything envisioned for the late twenty-fourth century, depending on how quickly this technology becomes implemented in a widespread fashion. The combination of motion sensors, facial recognition, software and hardware developments, and the price drop accompanying large-scale implementation could make the smart sliding door a routine part of our daily lives in the next decade. As far as the question of our technological capabilities, we're already there. The key is convincing the world that this technology belongs everywhere that humans go!

The Yorktown starbase provides artificial gravity using a complex system that results in a gravitational slipstream.

ARTIFICIAL GRAVITY

"It can be a challenge to feel grounded when even the gravity is artificial," Kirk once recorded in the captain's log.

Journeying through space, of course, there's no large gravitational mass to pull you toward it like on the surface of Earth. Instead, any gravitational fields will cause your spaceship to accelerate at the same rate as you and everything else on board. Effectively, without intervention, you would be weightless. Yet aboard every ship in the *Star Trek* universe, from the twenty-second century through the twenty-fourth, both crew members and inanimate objects experience the same gravitational force that one would experience here on Earth. Somehow, a mechanism had been devised to mimic Earth's gravitational force in space, leaving the space travelers of *Star Trek* with a very different experience from that of real-life astronauts of the twentieth and twenty-first centuries.

Artificial-gravity systems are fundamental to all ships in *Star Trek*, with the understanding that even if all other systems failed, artificial gravity would be perhaps the last to go. The technology is alleged to rely on a special type of gravity plating that could be applied to a ship's external hull or to the spaces between the interior decks. The plating is adjustable and customizable, so that something as large as a main deck or as small as an individual crew member's quarters can be tuned to their gravitational preferences. Individuals from planets with naturally lower or higher gravity than Earth might be at a disadvantage when gravitation was set to the "default" setting of Earth's gravity, but with the proper artificial-gravity settings, they could thrive in as natural an environment as possible.

As envisioned, gravity plating would confine the artificial gravity to a very narrow region of the ship: the interior decks. Anyone on the exterior would be effectively weightless, subject only to the accelerations of the ship itself, while anyone inside would experience the artificial gravity at whatever level it was set to in their location. There were some technical limitations that took quite some time to be overcome—in twenty-second-century starships, for example, there was an unavoidable "sweet spot" where artificial gravity was reversed from the rest of the ship, causing the forces along the vertical direction to become inverted.

According to our modern understanding of physics in general and gravitation in particular, there's no material that can create artificial gravity the way that *Star Trek*'s gravity plating can. Materials exert a gravitational force solely in proportion to their "gravitational charge," which we know as mass. While negative masses are a theoretical possibility—and negative mass or negative energy is a necessity for warp drive—there's no way to manipulate or tune a gravitational field in a region of space without their existence. Just as two types of electric charge, positive and negative, are required to set up a uniformly-directed electric field, the same principle holds true for a configurable gravitational field. Additionally, much more energy is required, since the gravitational force is inherently a factor of around 10^{38} weaker than the electric force between two protons. Hypothetically, one could manipulate individual gravitons as well, although no viable physical mechanism that would enable that has even been theorized.

But there's another approach that would work just as well and doesn't require any new physics beyond what Einstein realized more than a century ago: accelerating the ship. If you had a preferred direction for which you wanted to create artificial gravity—that is, a particular direction you'd want to be "down"—all you'd need to do is accelerate your ship in the opposite direction. To mimic Earth's gravity, you would apply a "gravitational thrust" that increased your ship's speed by 9.8 meters per second every second of your journey. You could adjust this artificial gravity to be stronger or weaker than Earth's by increasing or decreasing your rate of acceleration. According to Einstein's equivalence principle, to someone inside an isolated system where objects accelerate in a particular direction, there's no experiment you can perform that would differentiate between a force that arose due to gravitational acceleration from one due to accelerated motion.

The crux of this—and the reason it's called the equivalence principle—is because anything with mass responds to gravity and accelerated motion exactly the same way. It's an "equivalence" between gravitational mass (the mass that gets accelerated under the law of gravity) with inertial mass (the m in Newton's second law, $F = ma$.) By simply accelerating, everything on board will inertially try to remain in place, acting as an object at rest and exercising its Newtonian tendency to stay at rest. But the

While there is no preferred direction of "up" or "down" aboard the International Space Station due to the effective weightlessness and zero-gravity environment—as shown by Expedition 42 Flight Engineers Terry Virts of NASA and Samantha Cristoforetti of the European Space Agency (ESA), pictured aboard the ISS—the artificial gravity of *Star Trek* provides a downward pull on every deck that's indistinguishable from Earth's gravity.

accelerating ship around it causes the ship itself to exert the accelerating force, regardless of whether the ship accelerates in a straight line or in a circular fashion. In both cases, the sensation of artificial gravity will ensue, with the only difference being which surface appears to serve as "down."

In other words, fancy artificial-gravity generating technology isn't even necessary; all you need is an engine and you're guaranteed to experience it. Yes, there are downsides to generating artificial gravity through acceleration. A tremendous amount of fuel is necessary to continuously accelerate something as massive as a starship. And turning your engines off, down, or up will cause your artificial gravity to disappear, lessen, or increase, respectively, but that effect could frequently be used to your advantage. While your ship coasts at impulse or even warp speed toward your destination, traversing hundreds of millions of meters (or

While placing opposite electric charges on two parallel plates could create a uniform electric field, it would require a network of opposite gravitational charges—with positive and negative mass/energy—to create a uniform gravitational field between a ceiling and floor. Here, ISS astronauts Robert Curbeam and Christer Fuglesang construct a section of the Integrated Truss Structure during a 2006 spacewalk.

The combination of space suits, weapons, and magnetic boots were sufficient to defeat even the resourceful Borg in the zero-gravity environment on the main deflector dish outside of the *Enterprise*-E.

more) each second, a tiny, additional acceleration of 9.8 meters per second squared would supply you with all the artificial gravity you'd need. It wouldn't be tunable between individual decks or rooms like *Star Trek* envisioned, but it would provide the essential environmental force that human bodies require. It would eliminate the problems plaguing astronauts today, such as the deterioration of strong bones, severe muscular atrophy (including atrophy of the heart), vision and balance disorders, and the myriad of other maladies that come along with long-term weightlessness.

While the gravity plating of *Star Trek* might never come to be, the simpler and much more easily controlled artificial gravity of simple inertial acceleration will arrive as soon as humanity invests in building it. That technology and know-how is already here, and will be virtually indistinguishable from a planet's gravity once it's implemented. (The lone difference will be the absence of the minuscule Coriolis effect, which arises from Earth's rotation.) While there may be other ways to create a sense of "down" and "up," such as through the use of magnetic boots, they're not long-term solutions, as prolonged weightlessness is too detrimental to the human body. Still, if the ship's engines are completely off and it isn't accelerating due to any forces other than the outside gravitation of the universe, or if crew members need to perform a spacewalk, it might be nice to have a backup plan.

MEDICAL AND

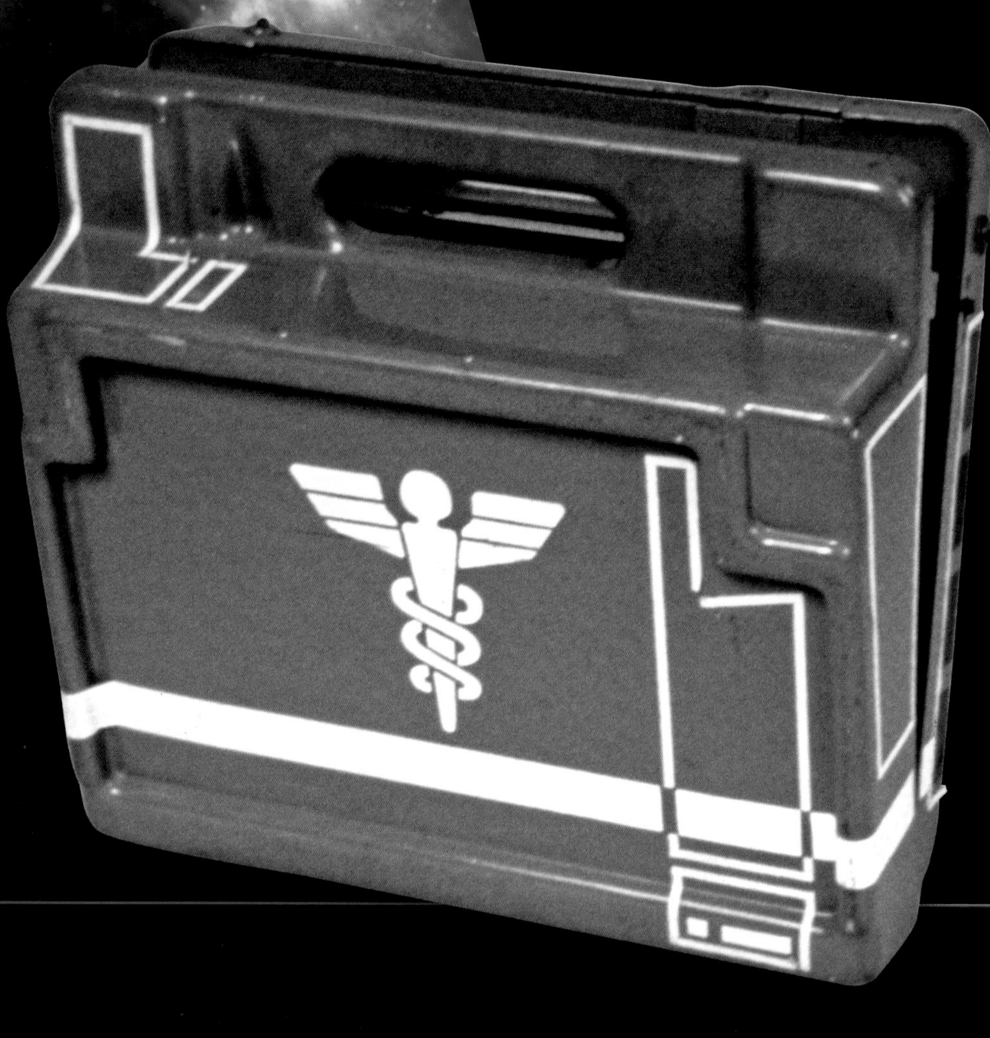

IOLOGICAL

Each of us, when we come into this world, is given one body to sustain us throughout our lives. Over time, we'll meet with disease, illness, and injury, which we'll do our best to fight through, mitigating and managing as best we can. Modern medicine and public health have come a long way, particularly over the past century, in increasing the average human lifespan well past seventy years of age, with many of us poised to enjoy a high quality of life even in our later years. But *Star Trek* envisioned a future in which the promise of a human life was far more extravagant: illnesses are cured quickly and easily, human lifespans routinely exceed one hundred years, medical maladies are diagnosed and treated without invasive surgeries, and conditions such as blindness represent an opportunity for the afflicted to be granted vision that exceeds the limits of human biology.

One of the hallmarks of modern medicine is the unfortunate fact that an invasive pathogen or tumor must be attacked, and that practically all methods of attacking it also damage the host's cells. Surgery, radiation, and treatments like chemotherapy are all incredibly harsh on the person eceiving medical care, no matter how well-localized the treatment is. But in *Star Trek*'s universe, surgery can be performed microscopically, without ever opening up the patient, hyposprays replace needles for injections, and tricorders can scan a patient remotely, obtaining all the information a doctor could want without even having to touch the patient.

Medical care in the real-life twenty-first century is going down this exact path. Greater numbers of mmunizations continue to render new, large classes of diseases irrelevant; surgical assistant robots enable more surgeries to be performed with smaller, less invasive incisions and faster recovery times; greater amounts of information can be gathered with less invasive techniques and blood draws; and human life expectancy and the time period for which we can expect to enjoy a high quality of life continues to ncrease. As scanning tools continue to improve in the era of "big data," preventive and personalized medicine are expected to increase life expectancies and quality of life even further.

Without a bona fide doctor on board, *Voyager* had no recourse but to use the only entity on board with medical knowledge and training: the Emergency Medical Hologram.

EMERGENCY MEDICAL HOLOGRAMS

"Please state the nature of the medical emergency." Familiar words in the late twenty-fourth century, as the Emergency Medical Hologram materializes into existence, prepared to serve as a medical expert in any situation in which a Starfleet ship needs an additional doctor. If too many crew members were injured in an accident, the ability to add capable, competent personnel at will with the most advanced training, skills, and knowledge available is a supremely beneficial option to have at one's disposal. In disaster situations where it was too dangerous for a human to venture, a hologram can rush to treat patients without worry for its own health or safety . And if a ship's only doctor takes ill or requires surgery, being able to call upon this resource to deliver top-notch medical care is invaluable. In the case of *Star Trek: Voyager*, the Emergency Medical Hologram was immediately and effectively put to use in a scenario it was never designed for: to serve as the ship's *only* doctor in all circumstances.

The creator of the EMH, Lewis Zimmerman, had some very strong opinions about medicine and a tremendous amount of knowledge, but that was dwarfed by the amount of information contained in the EMH program itself.

While Lewis Zimmerman might have trialed a human doctor—Julian Bashir—to be a template for a permanent medical hologram, robots are a much more practical solution given our current technology.

Artificial doctors—or nonliving replacements for other skilled human professions—are a long-running presence in the *Star Trek* universe. The first artificial unit to replace humans was all the way back in *The Original Series*, when Richard Daystrom's M-5 unit attempted to automate the functions of a captain and a majority of a ship's crew, running completely autonomously—and, infamously, failing the test run with fatal results. Prior to the M-5's disastrous failure, Spock noted to Bones that it was unfortunate computers had not advanced to the point where they could replace human surgeons. This might be a desirable end for anyone who isn't themselves a doctor—after all, a 2016 study revealed that medical errors may cause as many as 250,000 deaths per year in the United States, deaths that could perhaps be avoided if a computer were doing the work. The dream of a robot, hologram, or other automaton operating under its own power and programming might remove the great danger of human error, while providing the same high-quality treatment many of us expect (or hope for) from our medical care. With an incredible ability to hold more treatments, medical references, experiences, and memories than any human brain, Lewis Zimmerman's Emergency Medical Hologram represented a realization of this dream for *Star Trek* fans.

Fujitsu's ENON robot is fully autonomous, although it can perform only a limited range of functions.

While we're still a long way away from creating a tangible hologram of a person (see page 114), creating an automaton that allows a nonliving proxy for a doctor to perform medical duties is something that's well on its way. As the number of elderly citizens in many countries continues to rise—in Japan, for example, nearly 30% of the population is over sixty-five, and that's anticipated to rise to 40% by 2050—nonhuman medical personnel capable of administering care and treatment is fast becoming an absolute necessity. It doesn't look like the future that Lewis Zimmerman imagined, where a human would be used as a template for an artificial doctor, but on Earth in the twenty-first century, this is where machines come in.

By designing robots to accomplish specific medical tasks, various laboratories around the world, predominantly in Japan, have made tremendous strides toward developing robotic medical assistants. RIBA, the Robot for Interactive Body Assistance, was developed in 2009, and can lift a human from a wheelchair into a bed and back again, using its sensors to make sure that no one is dropped and that the transfer is comfortable. To avoid the phenomenon of the uncanny valley, which causes people to be unsettled by humanoid creatures that look so human that their unhuman features becoming unnervingly apparent, RIBA was designed not to look human, instead taking the form of a cartoonish bear. However, some scientists and engineers are working to overcome the uncanny valley and create a robot that appears truly human. Hiroshi Ishiguro at Japan's Osaka University has been one of the leading proponents of developing humanlike robots that can blink, speak, and breathe. His laboratory's Actroid line of robots, introduced in 2003, can be programmed in many ways and may someday overcome this limitation.

While these robots are either semiautonomous (like RIBA) or require a human operator (like the Actroids), there are also fully autonomous, programmable robots that have been developed. Fujitsu's ENON robot is self-guiding, programmable, and capable of speech synthesis and response. If a human-looking and human-acting robot can be programmed to interact under its own power, to make ethical decisions the same way a medical professional would, and to make general decisions in the absence of complete information (to extrapolate based on experience) the way a human would, we may truly find ourselves in the era of robotic medicine. Rudimentary robotic caregivers are already assisting patients in keeping to their medicine-taking schedules, taking basic vital statistics, looking for aberrations in their behavior, and assisting with basic tasks in the case of cognitive or physical impairment, such as the American-built Pearl the Nursebot.

In the meantime, robotic surgical assistants have already become commonplace. The precision with which machines like the da Vinci Surgical System can operate on a patient is so much more advanced than what a human being can ideally accomplish that it's becoming standard in a great many

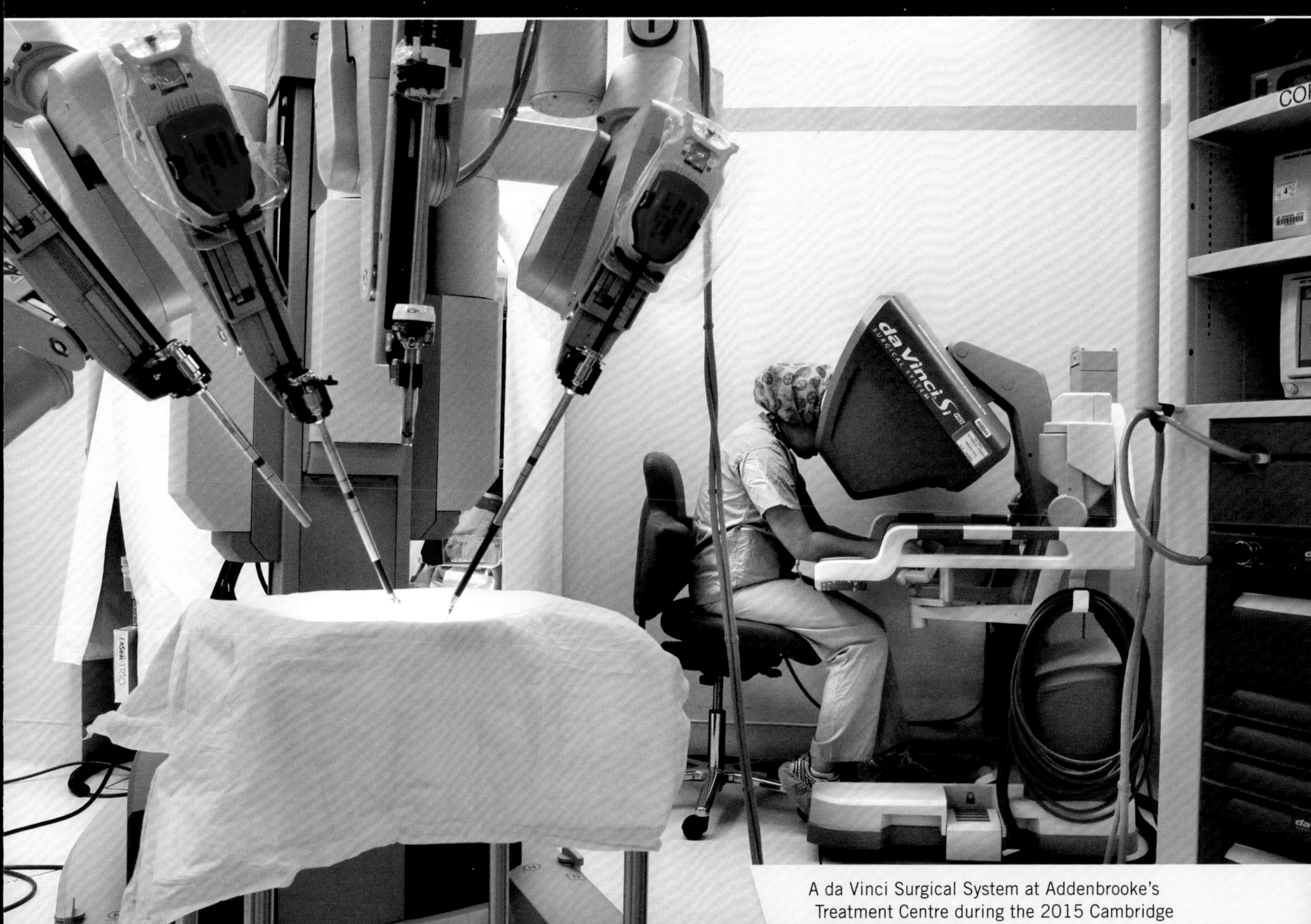

A da Vinci Surgical System at Addenbrooke's Treatment Centre during the 2015 Cambridge Science Festival.

hospitals for a variety of medical procedures. While onsite telemanipulation—or remote control from a human-driven input—was initially the most common way of operating such a device, the development of computer-controlled systems frees the human surgeon from needing to be located with the patient or device. Realistically, a surgeon can now perform a procedure with such a surgical system from anywhere in the world. The advances in this area have been tremendous on a number of patient-facing fronts, including precision of retractions, smaller incisions, decreases in blood loss, less pain, and quicker healing times. The patient outcomes for a great many classes of surgeries have improved as well, including reduced hospital stays, fewer transfusions, and reduced post-surgery pain medication.

One thing that *Star Trek* was successfully able to predict is the very large number of medical procedures that can now be performed without open surgery, which is a huge advance for both doctors and patients. Since the Hippocratic oath, the mantra to do no harm to a patient, is still followed closely in Starfleet, it's a tremendous boon to have one of the greatest risks to a patient—long-duration open surgical wounds—removed entirely.

The ability to create a true autonomous replacement for a doctor, nurse, or other medical professional may still be a long way away, but as time goes on, advancements in computing power, recorded experiences, and the ability to make decisions given limited information may someday allow surgical robots to perform surgeries *without* the assistance of a human doctor. In fact, it's not too difficult to imagine a future in which robots can perform those surgeries (and make those decisions) in a fashion that's superior to what a human could do! But despite the promises of big data, predictive analytics, and computational power, there's still no substitute for a human medical professional with training, experience, and compassion. The first replacements for human doctors are plausible given how technology is developing, but they likely won't look anything like the EMH Mark I from *Star Trek*. Perhaps other options—such as autonomous, specialized robots—will bring this emerging technology to fruition by the time the century is out.

CYBERNETIC IMPLANTS

Just like any other flesh-and-blood species, human beings are fundamentally limited in what their bodies can accomplish in any biological sense. Our bones, muscles, sensory organs, and neurobiology can only withstand and perceive external influences over a specific, limited range to which we became well adapted over millions of years of our evolutionary history. Yet despite these limitations, permanent human augmentation wasn't very common even in *Star Trek* unless there was some sort of handicap or birth defect that needed to be overcome. While Geordi's VISOR [see page 192] may be the classic example of human augmentation through technology in Starfleet, a very alien race—the Borg collective—made extensive use of cybernetic implants, enhancing the capabilities of all assimilated drones significantly over their purely biological origins.

When they first encountered the humanoid Borg drones, crew members of the *Enterprise*-D were shocked to discover how thoroughly augmented this race of creatures was. Although the physiology of individual drones varied tremendously, owing to the fact that they originally came from so many different species prior to their assimilation, they all received a large variety of similar augmentations. These

Before the *Enterprise*-D ever encountered the Borg, Magnus and Erin Hansen studied them for three years before being captured and assimilated. After assimilation, their daughter, Annika, was known as Seven of Nine.

cybernetic implants ranged from enhancements to wholesale replacements for various limbs and organs, including:

- homing devices and subspace transceivers, enabling them to communicate with the collective
- tubules and nanoprobes, enabling drones to assimilate others
- superhuman strength and power surge–resistant exoskeletons
- reinforcement of cardiopulmonary systems
- exoplating and a personal force field, rendering a drone resistant or even invulnerable against various physical and energy-based attacks
- eyepiece implants capable of recording the full spectrum of electromagnetic information about any and all objects, down to an object's molecular structure
- cortical implants of a wide variety, which allowed a drone's neural functions to interface with any external input desired

Assimilation of a non-Borg into the collective was a relatively quick process that took place through the injection of nanoprobes via assimilation tubules, with every known Borg drone possessing the capability of assimilating others. Captain Picard was known as Locutus during his assimilation by the Borg.

Borg drone has specialized cybernetic implants specific to its function and its species of origin, the chnology in any drone far surpasses anything available in the twenty-first century on Earth.

While the huge variety of potential cybernetic implants available prevents a comprehensive analysis of each one here, they fall into three general categories: passive internal implants, where a device is permanently inserted into the body to improve or regulate the function of a specific organ or system; active external augmentations, such as prosthetics, that can be controlled by the wearer's existing neurobiological system; and cortical implants, which couple an internal or external prosthetic or cybernetic enhancement with an enhancement or augmentation of the human brain and/or nervous system. At the time *Star Trek* simultaneously envisioned all three amalgamated together in the form of the Borg, only the most rudimentary implants and prosthetics were becoming available in the real world.

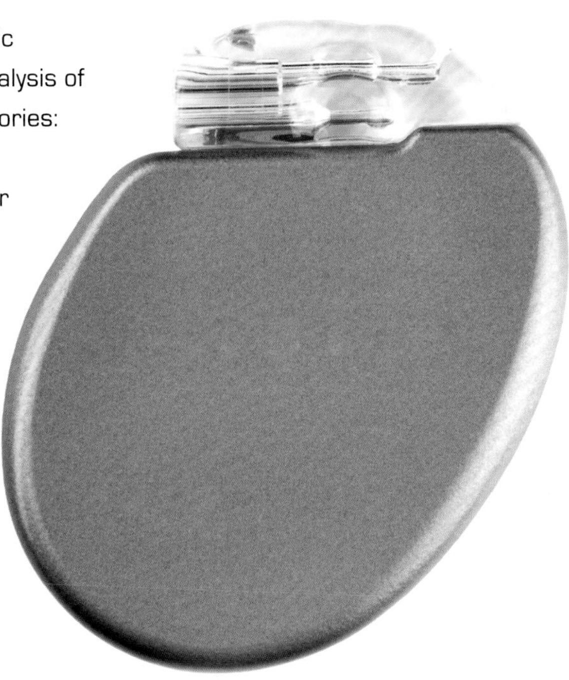

While pacemakers are the most common cybernetic implant in use, more advanced devices, such as the implantable cardioverter-defibrillator (ICD) shown here, can perform cardioversion and defibrillation as well.

Passive internal implants are common, and have been successfully implemented in a variety of formats. For people who've lost the ability to fully mobilize one or more of their limbs, joint replacement surgery has provided a huge improvement in quality of life that can last for decades. For diabetics, multiple painful needle sticks per day can now be replaced by a single implanted device that can be "recharged" by adding insulin every few days. And for those who have an arrhythmia (an irregular heartbeat) or other life-threatening heart condition, a variety of implantable devices that coax the heart into beating regularly are available, vastly improving the quality and span of life for millions of people worldwide.

External augmentations have undergone tremendous advances over the past few years as well, with neurally-controlled robotics rising to the forefront. In particular, myoelectric prosthetics, where an externally-powered artificial limb can be controlled by the electrical signals generated naturally by your own muscles, have gone mainstream in the 2010s. The residual limb can interface with a custom prosthetic socket, the strength and speed of the prosthetic's movements controlled by the electrical signals arising from the muscle intensity. Prosthetics are available for all four limbs with a variety of functions and controls to either reproduce or supersede the functions of a human hand, with 360-degree wrist motions built into many. Functions as delicate as cracking an egg, picking a grape, or pouring a beer into a glass have been demonstrated many times. Hybrid prostheses can combine myoelectric-controlled and body-powered components to control elbow, knee, shoulder, or hip functions.

But the most advanced cybernetic implant of all is one that would be wired directly into a human brain: a cortical implant. By allowing the mind to interface directly with cybernetic devices, even those with no muscular control at all could receive inputs from and deliver outputs to various external devices. Most importantly, it would restore the ability to communicate for patients who are aware but have no means of effectively interacting with the outside world. In 2016, the very first successful cortical implant into a patient with ALS was completed, marking the first fully implanted brain-computer interface in a human being. The recipient, through a series of implanted electrodes and transmitters, was able to select letters by attempting to move her (completely paralyzed) right hand, and the stimulated nerves sent the signal to the transmitter which then interfaces with the computer. Although it's slow going—taking twenty seconds per letter—it's a 95% accurate process and doesn't rely on eye-tracking or other technologies unavailable to those with total paralysis. Thanks to cortical implants, modern technology has demonstrated the ability to restore communication to those who've lost the ability.

As further neural pathways are mapped out, become controllable, and are tied to transmitters, computers, and interfaces, we can expect the process of communication to speed up tremendously, someday approaching the speed of sign language used by the deaf community. In addition, the first

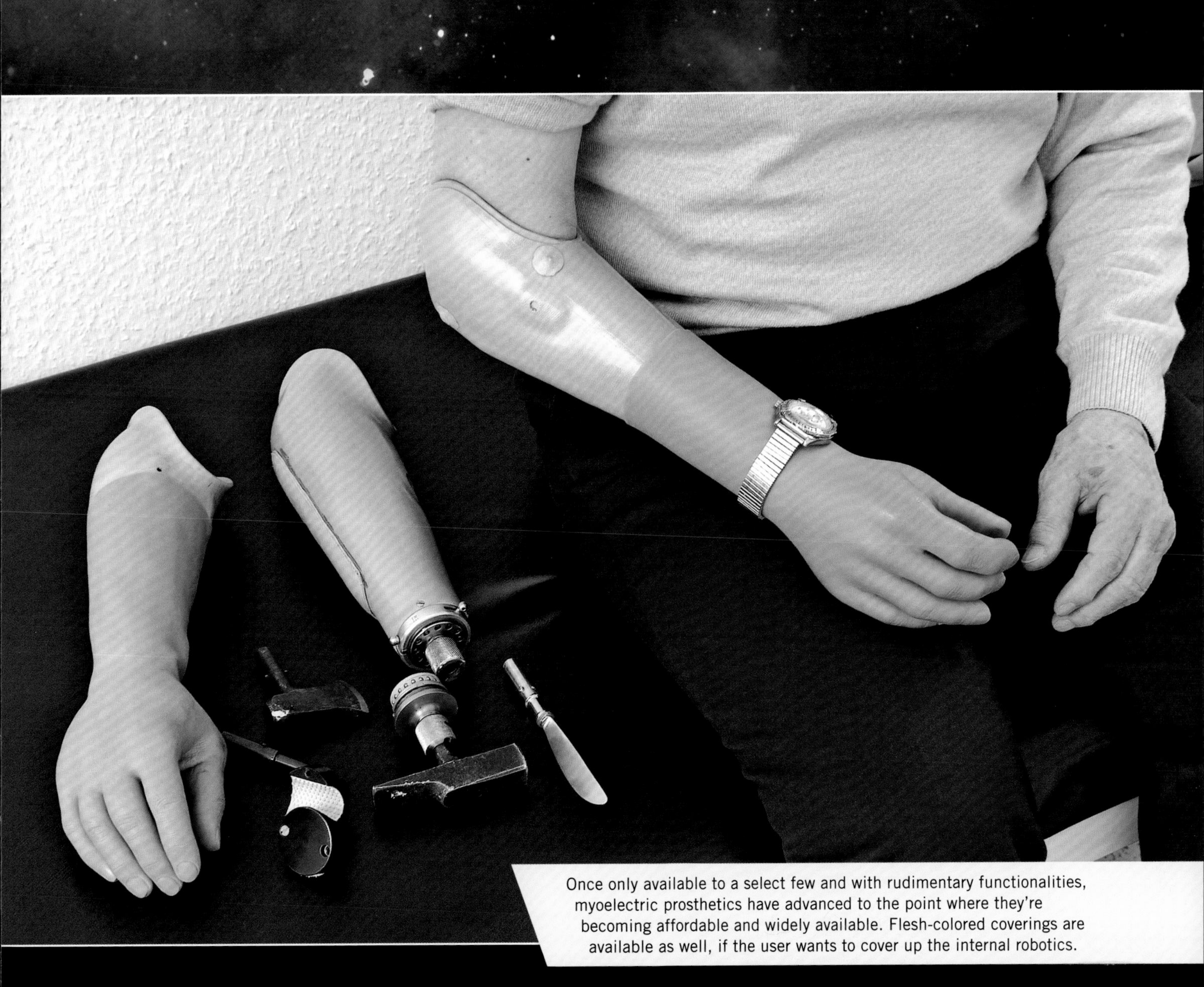

Once only available to a select few and with rudimentary functionalities, myoelectric prosthetics have advanced to the point where they're becoming affordable and widely available. Flesh-colored coverings are available as well, if the user wants to cover up the internal robotics.

Stephen Hawking, afflicted with arteriolateral sclerosis (ALS), can control a speech synthesizer using a cheek muscle over which he still has control. Many patients with ALS or locked-in syndrome cannot move any muscles, but advances in cortical implants and brain-interface technology may render communication possible for them after all.

successful cortical implant offers a window into the potential of this technology. With direct links to an external computer, perhaps electronic communication via simple mental controls—akin to telepathy—will become widespread. Perhaps we won't need computers or smartphones to send and receive email or messages; perhaps a simple set of implanted augmentations will make that possible. And for both internal and external cybernetic implants, perhaps restoring or recharging power and upgrading software will be able to happen simply by hooking ourselves up to the proper interfaces.

This brings up the classic philosophical question of where the line is drawn between an augmented human with cybernetic components and a true cyborg. It also brings up the ethical question of whether one can lose one's humanity by becoming more machine than human. How human was Seven of Nine when she was part of the Borg collective? How about Locutus? Or Hugh? And does being severed or plugged in change their nature? It brings up the simultaneously wonderful and terrifying possibility of being plugged into a network of all of humanity's thoughts as a real-life incarnation of the Borg collective. While cybernetic implants are taking off on multiple fronts to aid and empower those of us who are ailing, the possibility of augmenting our lives is a tantalizing one. As this technology continues to develop, many new possibilities for what humans are capable of will continue to emerge. May we use this technology responsibly, and may it never fall into the hands of those who would use it to hack our minds or bodies to take away our own bodily and mental autonomy.

Admiral McCoy took a tour and conducted an inspection of the *Enterprise*-D, telling Lieutenant Commander Data, "Treat her like a lady and she'll always bring you home."

HUMAN LIFE EXTENSION

"He was the best of the tyrants and the most dangerous," James Kirk said of Khan Noonien Singh after their first encounter. As an Augment—a human created to be exceptional through genetic engineering and DNA resequencing—Khan possessed five times the strength and double the intelligence of an average human as well as enhanced senses, immune systems, agility, and heart and lung capacity. But perhaps most interesting was the idea that Augments had double the lifespan of a normal human—that simply altering a human's genetic code could give them an extra eighty to one hundred years of life. Although the development of the Augments was, according to the *Star Trek* timeline, supposed to have happened in the late twentieth century, it would take centuries to bring about the much hoped for gradual improvements in human longevity. By the twenty-second and twenty-third centuries, the mean human lifespan was around 100 years, and by the twenty-fourth, a 120-year lifespan was considered normal. Leonard McCoy from the original *U.S.S. Enterprise* was present for the maiden voyage of the *Enterprise*-D, at the age of 137.

If we could remain healthy, vital, strong, and mentally sharp, wouldn't we want to live as long as possible? While healthy lifestyles with proper diet and exercise combined with cures for a variety of diseases, ailments, and sicknesses might have added decades to the average human lifespan over the past two centuries, we're fast approaching the limits of what it can accomplish. Genetic engineering, like that responsible for the Augments, holds the greatest promise for extending the human lifespan. But the ethical questions are tremendous—what consequences would we face if we chose to engineer "designer" human beings? While few would argue that gene editing to cure deadly genetic diseases is unethical, selecting for or against various traits could be severely detrimental to the genetic diversity that humanity presently enjoys. In *Star Trek*, the Augments are shown to possess—along with their positive superhuman traits—a short temper, a propensity toward egomania and violence, and a diminished sense of what we commonly understand as morality. When the Augment Malik catches Dr. Arik Soong tinkering with an embryo, attempting to alter the genetic propensities he saw as negatives, Malik chides him: "You're changing its personality." Soong retorts, "I'm correcting a defect in its genome." (The disagreement would lead Malik to imprison Soong and, later, attempt to murder him.)

Artificially selecting for a particular DNA sequence by necessity means that alternative sequences are selected against. Even benevolent attempts at genetically modifying human beings may hold unintended, negative consequences.

Even though the human genome has been sequenced, it's unclear which genetic changes will result in multiple, perhaps unforeseen, phylogenic changes in a human's development.

Today's best science indicates that combating disease and malnutrition while simultaneously living a healthy, active lifestyle should get most human beings to the age of eighty, and a small percentage of those will live to see one hundred. While the developments of germ theory, antibiotics, antiseptics, and modern medicine have improved our prospects tremendously over what they were just one hundred years ago, these advancements aren't without bounds, and we seem to be running up against the limits of what lifestyle changes can bring us without further medical or biological advances. As we age, our immune systems begin to fail us more and more frequently. A variety of cellular and humoral immune responses are seen to change, with altered T-cell phenotype and reduced T-cell response—making our bodies slower to react to and less effective in fighting foreign pathogens—being perhaps the deadliest changes observed.

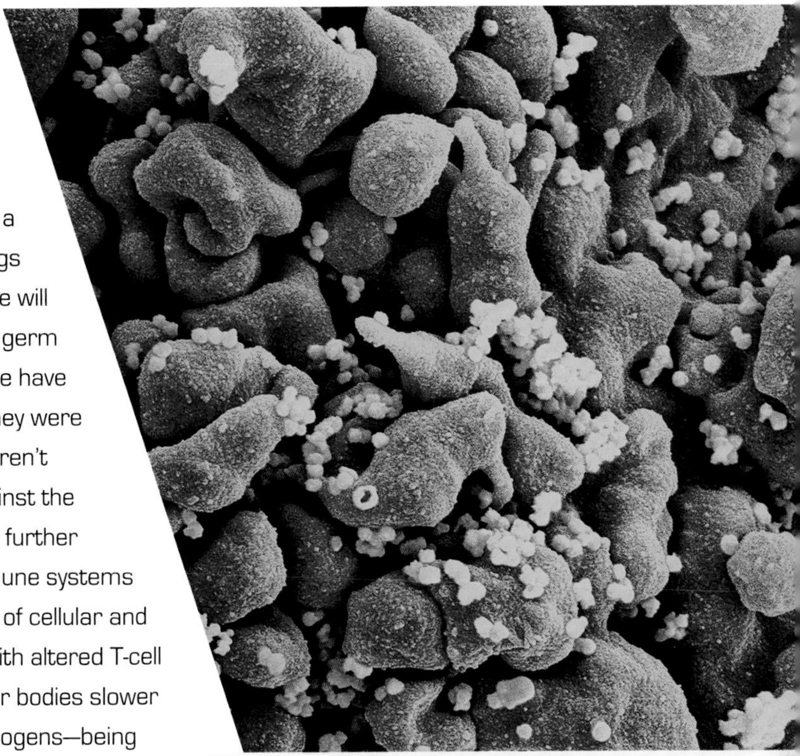

An infected T-cell cannot perform its immune system functions properly, while producing the wrong types of T-cells prevent a body from combatting infections in an effective manner.

Perhaps the ultimate goal of human life extension research is to discover what the full suite of changes are that aging induces, and to engineer a solution to keep us effectively ageless. Today, a twenty-year-old woman in the United States has only a 1 in 2,000 chance of dying in the next year; at forty, she has a 1 in 500 risk; at sixty, it's 1 in 120; at eighty, 1 in 20; and by ninety-five, the odds are 2 in 3 that she won't make it to ninety-six. If we could keep our bodies functioning at the same level at which a young, healthy person's body functions, could we far exceed even the dreams of life extension exalted by *Star Trek*? In the modern era, Jeanne Calment made it further than any other human has, having sold pencils to Vincent van Gogh as a child and eventually dying in 1997 at the age of 122. Yet all attempts at lifestyle modification to increase longevity—whether by supplementation, diet and nutrition, calorie restriction, or prescribed activity levels—have failed to push any substantial fraction of the human population above the hundred-year mark in terms of life expectancy. Proteins like GDF-11, stem cell research, supplements like Elysium Health's Basis, hormone treatments, and others are all presently being tested. These upper limits to human life expectancy have failed to advance for multiple generations, however; for the time being, extending our lifespans by conventional means appears to have reached its limits.

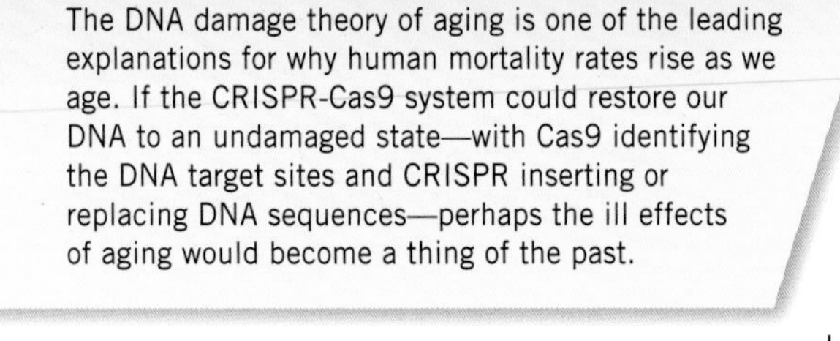

The DNA damage theory of aging is one of the leading explanations for why human mortality rates rise as we age. If the CRISPR-Cas9 system could restore our DNA to an undamaged state—with Cas9 identifying the DNA target sites and CRISPR inserting or replacing DNA sequences—perhaps the ill effects of aging would become a thing of the past.

The greatest hope of human life extension doesn't come from hormone replacements, blood transfusion, or embryonic gene modification, but rather from the strategic use of prokaryotic DNA segments known as CRISPR. CRISPR, short for Clustered Regularly Interspaced Short Palindromic Repeats, has enabled the most remarkable biotechnology imaginable for keeping an organism like a human being alive: genome editing inside a living being. The CRISPR system works in primitive organisms by saving a piece of foreign DNA from an invading organism, inserting it into its own DNA (identifying the appropriate site using the protein Cas9), and then being able to read that DNA in the event that a similar-enough organism invades again in the future. But genome editing isn't limited to saving bits of DNA from the organisms that try to kill us; the CRISPR-Cas9 system is *programmable*. Unlike traditional genetic engineering, which required a new generation to

While it's easy to foresee the positive consequences that making any particular genetic modification might have, the negative ones may prove more difficult to envision.

While we might be able to engineer humans to become centenarians with near certainty, perhaps even maintaining an incredible amount of vitality, the possibility of halting the aging process altogether holds an even greater allure than *Star Trek* ever envisioned.

make changes in an organism's genome, CRISPR can edit living cells, switching genes on and off, inserting sequences, or even rejuvenating damaged DNA. CRISPR may have the potential to defeat retroviruses such as HIV or to cause your immune system to better identify and kill cancer cells—both applications for which it's presently being used. There are thousands of diseases caused by genetic imperfections caused by a single letter-error in human DNA; CRISPR could perhaps cure all of these. Aging, from a DNA point of view, is a disease in its own right. A 2014 meta-analysis of thirty-six studies involving nearly five thousand individuals showed an extremely strong correlation between age and DNA damage. Could the CRISPR-Cas9 system undo this damage to the human DNA contained within our cells? And, if so, could that cause our mortality rates to plummet to the same levels they were at when we were only in our twenties?

This is one of a number of emerging technologies that enhance our ability to do genome editing for either current or future generations of humans, dependent on which cells are edited and at what stage of a human's life. Just as the science (or, perhaps, the pseudoscience) of eugenics was determined to be discriminatory and unethical, in violation of human rights, perhaps genome editing for future generations of humans contains too great a potential for abuse. Perhaps whatever guidelines or policies for use are put in place, they will be subject to the whims of whatever group is in position to make those guidelines, or perhaps the loss of genetic diversity will be a greater loss than any gains an individual might make. The ethical questions raised by *Star Trek* some fifty years ago, when the Eugenics Wars and the Augments were conceived, may in a sense be more real today than they ever were when Khan made his first appearance.

At the time *Star Trek* premiered, the DNA molecule governing inheritance had only had its chemical structure discovered thirteen years prior, in 1953. On the other hand, Mendelian genetics was almost ninety years old, and the theory of genetic inheritance had been around for many decades, along with the disastrous eugenics movement of Nazi Germany. It wasn't unreasonable to imagine that health and medical improvements might lead to "normal" lifespans of a hundred years, and that embryonic genetic modifications might even be able to double that. But no one could have fathomed that in the early twenty-first century, we'd be performing precision gene editing on a genome more than three billion base-pairs long, while the organism was still alive. We don't have the solution to aging yet, but the technology that may deliver it is already here, and its full potential is more powerful than anything *Star Trek* ever dreamed up!

Geordi La Forge, with his VISOR and with his later ocular implants.

VISORs

"You know, I've never seen a sunrise . . . at least, not the way you see them," says Geordi La Forge to Captain Picard. Like so many people before him, from either birth or a young age, Geordi appeared destined to live his entire life without his sense of sight. But in the twenty-fourth century, technology had advanced to the point that, despite the *Enterprise*-D engineer not having a working connection between his eyes and his brain via the optic nerves, a prosthetic device known as a VISOR (for Visual Instrument and Sensory Organ Replacement) could overcome those limitations. The VISOR can not only transmit external visual information to his mind but show him the universe far beyond what human eyes can see. While we might know only a tiny fraction of the electromagnetic spectrum, visually, the VISOR enables the wearer, through a direct link to the optic nerve, to process information ranging from radio waves all the way up to ultraviolet light.

We often view the loss of one of our senses as a handicap in this world, but in the universe of *Star Trek*, it opened up an opportunity to go far beyond our biological limitations. More than merely restoring perception of normal human wavelengths to someone with a visual impairment, an external prosthetic could detect signals covering the entire electromagnetic spectrum between 1 and 10^{17} Hz. These signals could then be transmitted through the optic nerve to the brain directly through neural implants wired to the wearer's temples. While the VISOR's information was confusing and disorienting to humanoid eyes, an experienced wearer could discern things that were completely opaque to a biologically-sighted human. Fluctuations in body temperature, perspiration levels, and other biological signals could be detected just as easily as the color of someone's shirt, granting the wearer the ability to perform limited medical diagnostics—or to detect the signals of a lying companion—from afar.

Many people struggle with vision loss or vision impairment in a variety of forms. Up until recently, the only options were corrective lenses or surgery, and even those options were severely limited in terms of the types of impairments they could correct for. The way your eye works, from a simplified perspective, is that incoming light enters through your pupil, is focused by your eye's lens, travels through your vitreous fluid, and strikes the rods and cones on your retina. That information is then transmitted to your brain by the optic nerve, which transforms that signal into an image, enabling you to see. The most common correction that can be made for someone with imperfect vision is the simple addition of an external lens, either through glasses, contact lenses, or laser eye surgeries like LASIK or PRK. Corneal transplants are also common, and more recently, experimental retinal transplants have been performed. In 2014, the first retinas grown from a patient's own stem cells were successfully transplanted into her eyes, giving her sight again after macular degeneration—the most common cause of blindness in humans—had taken most of it away.

But for those who suffer from incurable or noncorrectable vision impairment or even total blindness, technology options were few until the turn of the twenty-first century. In the late 1990s, NASA began working on devices that a vision-impaired patient could wear over their eyes and, used in conjunction with a computer, enlarged and brightened text, enabling them to perform tasks like reading small print and filling out forms. The result, debuting in 1999, was a low-vision headset that could be worn akin to Geordi La Forge's VISOR. The device, appropriately named the Joint Optical Reflective DisplaY (JORDY), can magnify objects up to a factor of thirty times when worn as a pair of glasses. The wearer can also adjust brightness, contrast, display modes, and more, enabling those with low vision to experience aspects of life that would be impossible without technological assistance. Further enhancements in the 2000s included image stabilization and the ability to focus on objects at any distance; wearers of JORDY can do everything from enjoying sporting events to watching television to reading small-print text to seeing the faces of their loved ones.

There are an estimated thirty-seven million people in the world who are blind for various reasons and to various degrees. For someone with an impairment like Geordi's—total blindness from birth—there was very little to be done at the time that *Star Trek: The Next Generation* premiered. The one glimmer of hope was from an experiment dating back to 1755 and the French physician and scientist Charles Leroy; by discharging an electric current through a blind patient's body, from above the eyes to below them, Leroy discovered he could induce a momentary visual signal. This rudimentary setup could be considered the first prosthetic visual device. When the light-receptor cells in the eyes degenerate, as from retinitis pigmentosa, or through damage to the optic nerve due to trauma or glaucoma, a visual prosthesis that bypasses the normal eye anatomy is the only option. By combining a video camera hooked up to a pair of prosthetic glasses with a retinal implant, external visual data can be converted

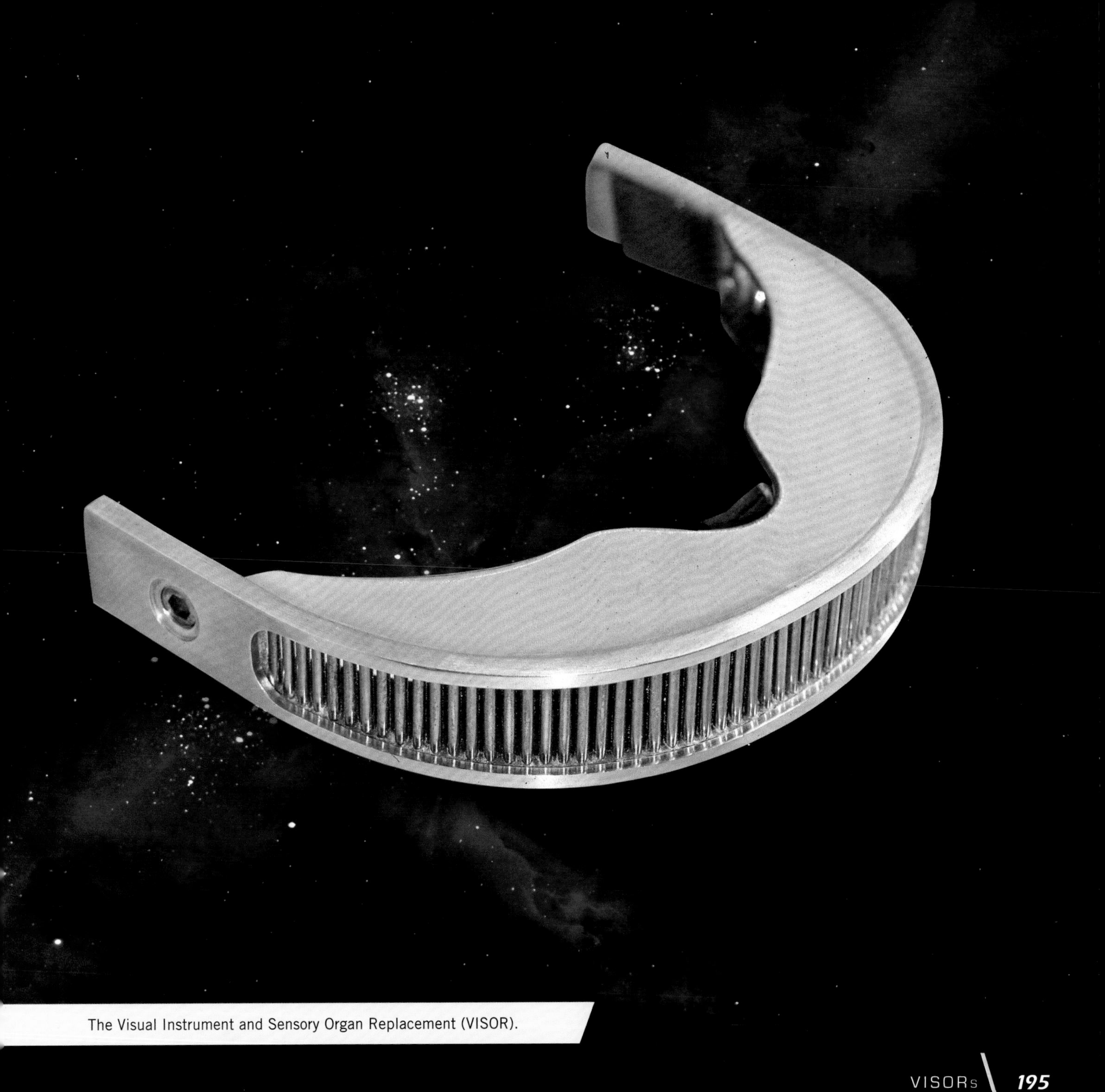

The Visual Instrument and Sensory Organ Replacement (VISOR).

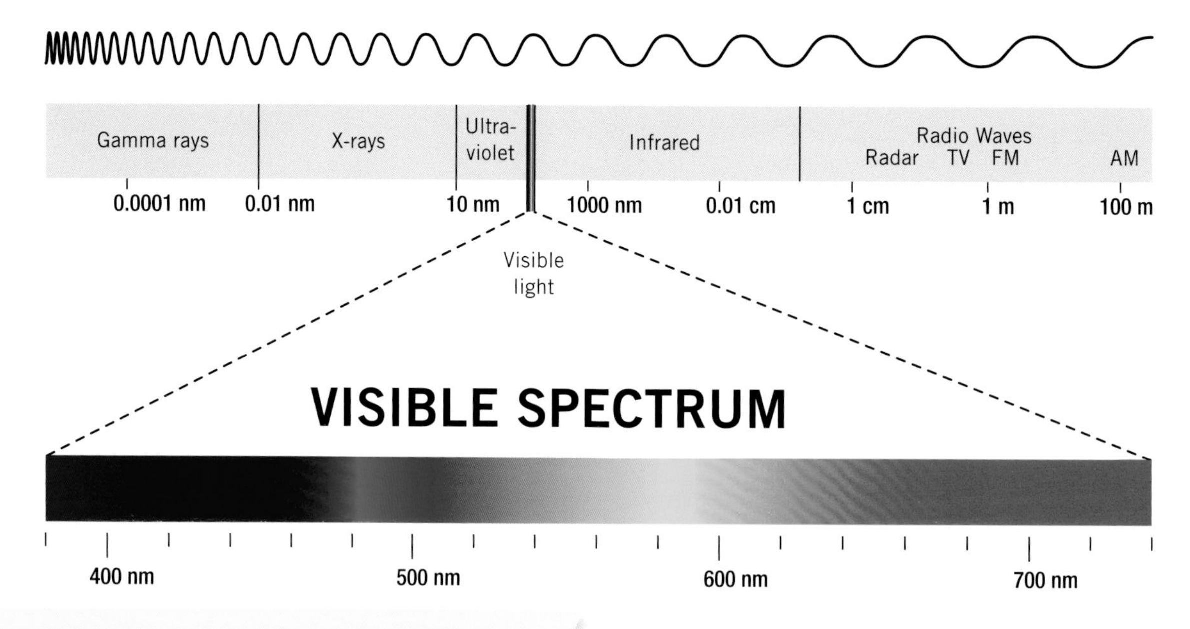

While human vision is limited to the range between 400 and 700 nanometers in wavelength, the VISOR was able to detect electromagnetic radiation over a range dozens of times as large.

into electrical impulses and transmitted through the optic nerve. Groups such as Bionic Vision Australia and the Boston Retinal Implant Project have prototype devices that do exactly this at present, but a group at Monash University superseded all of them in 2012: it built a working device to transmit optical information directly to the wearer's brain, even enabling those without eyes or optic nerves at all to receive visual sensory input. Combined with eye-movement sensing technology and wireless transmitters, the current prototypes are estimated to be of potential assistance to up to 85% of those currently afflicted with blindness.

Modern technology is a long way from transmitting all the detailed information that a multiwavelength camera would collect directly to the wearer's brain the way Geordi's VISOR does, but the proof-of-concept technology is already in use. In 2012, the retinal code used to send signals to the brain was cracked in mammals, literally restoring sight to blind mice. The ocular response has been successfully mimicked using glasses and cameras, and as the technology becomes approved for human trials, it seems only a matter of time before the visual equivalent of cochlear implants—ocular prosthetics—becomes as successful at restoring vision as *Star Trek* imagined. It also opens up the possibility that hackers will be able to tap into a blind person's device and see the world through their "eyes" in the ultimate act of voyeurism.

It seems odd that, despite Geordi having received his first VISOR at the age of five, neither Dr. Crusher nor Dr. Pulaski had ever seen one before meeting him. Although *Star Trek: The Next Generation* is set more than three hundred years in the future from our present day, the key breakthrough of cracking the retinal code for vision has already been achieved. By figuring out how to transmit the external information the world offers to the brain's visual cortex, it's quite conceivable that we'll soon be able to map electromagnetic information from far outside the visual range into false color, enabling humans with such a device to see it. In addition, as camera and computer technology continues to miniaturize, the VISOR may never need to become a widespread technology, as humanity might advance rapidly straight to ocular implants. At this point, it would be almost inconceivable if the vast majority of blind individuals didn't have most, all, or even more than 100% of their sight restored by the middle of the twenty-first century. This is another spectacular example where the science fiction predicted by *Star Trek* is coming to fruition faster than its creators ever imagined.

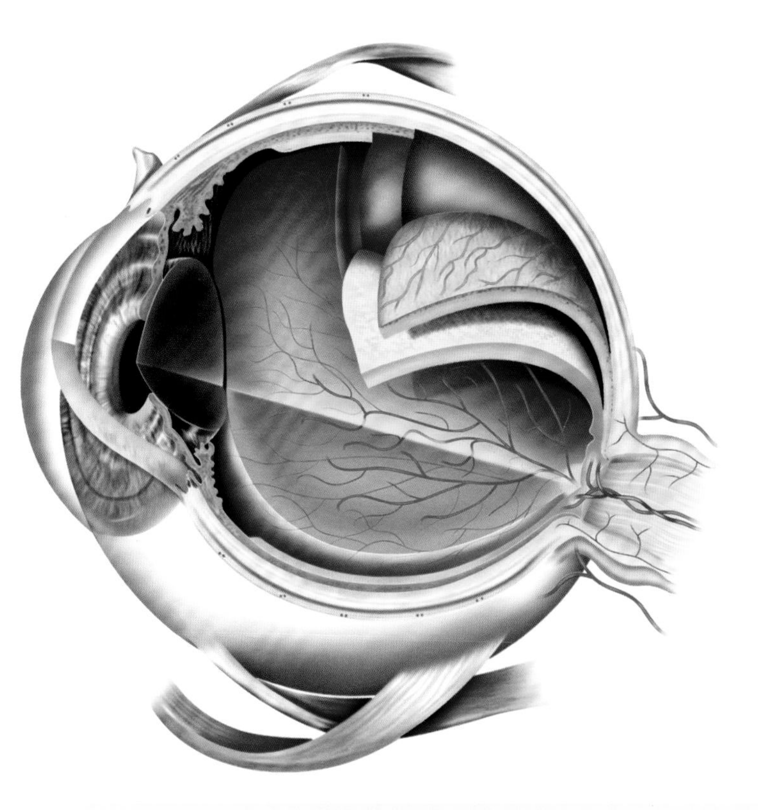

The anatomy of a human eye consists of many parts, but what gives you sight is the transmission of light through your pupil, focused by a lens onto the retina, where the signals are then transported to your brain via the optic nerve.

By bringing along a small complement of medicines, a doctor armed with just single hypospray device could administer whatever medicines were deemed necessary in a matter of seconds.

HYPOSPRAYS

"This won't hurt a bit, Spock," says McCoy, who's met by the rejoinder, "An unnecessary assurance, doctor, in addition to being untrue." While most of us have come to expect painful needle injections as a standard part of medicine, *Star Trek* offered the promise of a technology that could successfully deliver medicine to any desired injection site without any sharp objects at all. Instead, a spray of high-pressure air would penetrate through the skin—even through clothing, if necessary—administering the appropriate dose of liquid medicine into the patient's subcutaneous tissue, muscle, arteries, or even vital organs as required. The lack of a physical contact surface like a needle meant that the possibility of disease transmission was greatly reduced and that a single device could be used for an arbitrary number of patients with no waste, no risk of patient contamination, and no need for sterilization. In short, being injected with a hypospray offered the patient increased safety, reduced discomfort, and virtually eliminated their risk of infection.

Lawrence E. Blackman receives the first typhus shot administered by Hypospray Jet Injector in 1959.

Rather than taking pills or liquids orally, a hypospray could be safely and easily administered for even simple maladies, like nausea or gastrointestinal distress. Using a hypospray on a patient afflicted with a blood-borne illness would pose no threat to subsequent patients, even if a doctor were using the same hypospray with no sterilization procedures. The dose could be altered by simple controls on the head of the device itself, and injections directly into the carotid artery could provide nearly instantaneous relief by accessing the blood-brain barrier in mere seconds. The sound and a small sensation of pressure is all a patient would experience before feeling the effects of the medicine inside. This made the hypospray useful for not only medicinal purposes, but as the ultimate tool to stealthily incapacitate an unsuspecting foe.

While it may seem unique and innovative, the hypospray is one technology envisioned by *Star Trek* that actually existed prior to the show's premiere. Known as a "jet injector," it's powered by a high-pressure system of air or gas that allows a liquid dose to be administered in a stream, through the skin and into the patient. While accidental jet injections in humans have been around since the first high-powered grease guns were developed in the nineteenth century and became (unfortunately) commonplace with the popularity of the diesel engine in the early twentieth century, a purposeful jet injection held tremendous medical potential. For administering a large quantity of vaccines to massive numbers of patients both cheaply and rapidly, there was no better option.

Thanks to an advance made by Aaron Ismach, jet injectors not only became portable, but were modified so that they could administer liquid just barely beneath the skin: perfect for a smallpox vaccination. In 1967, just a year after *Star Trek*'s debut, the World Health Organization called for a massive vaccination campaign to rid the world of smallpox. Ismach's device was capable of rapidly vaccinating up to one thousand people per hour and was instrumental in vaccinating huge swaths of the African continent. Smallpox was officially announced to be eradicated in 1980.

Despite the jet injector's incredible success—it helped set a world immunization record in 1976 when fifty million Americans were vaccinated against the swine flu in only ten weeks—it's rarely used

Different models of hypospray devices.

The jet injector gun, one of Aaron Ismach's developments, has fallen out of favor from its heyday in the second half of the twentieth century and is rarely used any longer.

today for any purposes. While *Star Trek* might have envisioned that the hypospray would eliminate the risk of cross-patient contamination, real-life jet injectors suffered from a seemingly unavoidable problem: breaking the patient's skin resulted in a risk of getting their blood on the device. (Exactly how this was accomplished, however, has never been explicitly addressed in *Star Trek*, in any of its incarnations.) As the nozzle frequently became contaminated, the device was no longer deemed safe. Even adding on a single-use protector cap didn't solve the contamination problem, as infectious agents like the Hepatitis B virus were found to permeate the internal fluid administration pathway of the jet injector. After a number of these revelations, the U.S. Department of Defense announced it would stop using jet injection for mass vaccinations, effective as of 1997.

However, the discomfort and—for many—the fear of needles remains a problem for public health and medical professionals. Patient compliance, particularly for those required to receive large-dose or frequent injections, is an obstacle that must be overcome, and work continues on a viable hypospray. Single-use jet injectors for administering vaccines to young children and individual devices for administering insulin to diabetics are rare, but still available upon request. In 2012, the MIT BioInstrumentation Lab developed a new hypospray-like device that is completely programmable in terms of pressure. It's not only capable of delivering large, protein-based injections to patients, but can be tailored to ideally penetrate an adult's scar tissue just as easily as a baby's skin. This novel device injects material at the speed of sound, but its most impressive feature is that the injection cross-section is as small as a mosquito's proboscis, meaning it's completely painless.

Right now, many challenges currently facing doctors working to administer vaccines and medicines around the world could be solved by a successful hypospray. Medicines that need to be refrigerated could instead be dehydrated and powdered, eliminating the huge difficulty of keeping them consistently cooled. Injections that need to be administered subcutaneously, intravenously (or intra-arterially), or directly into a muscle or organ could be successfully sprayed into the patient by setting the injection pressure correctly. And rapid, reusable administering of such medicines could be accomplished without risk of contamination with high-powered versions of existing technology, such as by sterilizing the equipment between injections with the modern technique of ultraviolet germicidal irradiation. The idea of a single device that could be used to inject as many patients as needed without delay, risk of injury, contamination, or other adverse effects, and limited only by the dose of the medicine available, seems to easily be within reach of current technology. It's only the details that need to be worked out.

Unfortunately, one of the biggest obstacles envisioned to the administering of medicines in the twenty-second, twenty-third, or even twenty-fourth centuries is widespread in our twenty-first: patient noncompliance. While it's the great hope of many that the proven safety, quality, and efficacy of "taking your medicine" would outweigh any fears—rational or irrational—among potential patients, the reality is that people frequently opt out of treatments that could vastly improve their quality of life. There's no guarantee that even improving those aspects, along with eliminating the pain of needles, would do anything to change that.

As advanced as medicine may become, the idea of forcibly medicating (or performing medical tests on) noncompliant patient raises the types of ethical questions that *Star Trek* is so renowned for. If an individual, race, or species has wronged you in the past, but you are biologically capable of helping them survive, can you be compelled to assist? If there's a threat to your civilization, should mandatory testing or medication be enforced? Or is an individual's autonomy over their own body a more fundamental right—and to hell with the consequences? While neither modern humans on Earth nor future civilizations in *Star Trek* have arrived at an all-encompassing answer, perhaps the hypospray may someday play a role in making it easier for individuals to choose to benefit not only themselves, but society as a whole.

The version of the hypospray developed by the MIT BioInstrumentation Lab may be the first pain-free, pressure-customizable jet injector, with the potential to someday replace needles entirely.

TRICORDERS

"You're showing the same distorted readings," says Dr. Crusher to a patient who's been engaging in very suspicious activities. While the patient might lie to the doctor, data taken firsthand from a patient reveals the truth to a trained professional every time. Simply scan your tricorder over a patient's body, or over a particular body part you want to examine more closely, and you can instantaneously, remotely, and noninvasively learn a whole slew of information. What are his vital signs? How has her DNA changed since her last scan? What injuries or traumas has he suffered? Is she ill? Is any aspect of his blood chemistry out of the normal range? Has she contracted any diseases? And were you tinkering with that experimental warp device that we warned you about, Wesley?
The tricorder reveals all.

The original tricorders were large, boxy objects that performed various functions, and were essential equipment among science and medical personnel. When tailored for medical diagnostic purposes and outfitted with small, removable handheld scanners, they were the first tools used in information-gathering about a patient. They could scan an entire patient or just focus on a specialized area, and became progressively more advanced as *Star Trek* progressed from the twenty-second to the twenty-third to the twenty-fourth centuries. They were capable of performing a full blood analysis on a patient and enabled medical staff and officers to scan and detect abnormalities without the full suite of equipment available in a sickbay facilities. On an away team mission, the tricorder was practically as vital a piece of equipment as a communicator (see page 94) or a phaser (page 68).

Although tricorders varied among empires, in shape, size, and functionality and across time periods, they were all handheld devices capable of gathering large amounts of data and performing a variety of tests and functions remotely. Many had specialized functions, but the device is most famous for its medical use.

Tricorders could sense and detect a huge variety of physical, chemical, and biological signatures, and the complementary scanner enabled very particular locations or body parts to be examined in even greater detail. While tricorders could be specialized to a great many functionalities, including as gas-and-air analyzers, X-ray machines, multispectral imaging devices, remote transmitters, and more, one of their best-known uses has been as multifunctional medical scanners. Rather than a host of equipment and laboratory tests needed to examine a patient like we do today, *Star Trek* envisioned a future where just a single motion of a scanner coupled to a tricorder could detect all sorts of medical and biological properties about a patient. One doesn't even need to be looking for the right thing, necessarily, to have the tricorder's analysis reveal the incontrovertible truth about what's happened to a patient's body.

There are a large number of things, physically, that can most definitely be measured remotely, simply by knowing where to look. The body heat people give off can be used to measure their internal

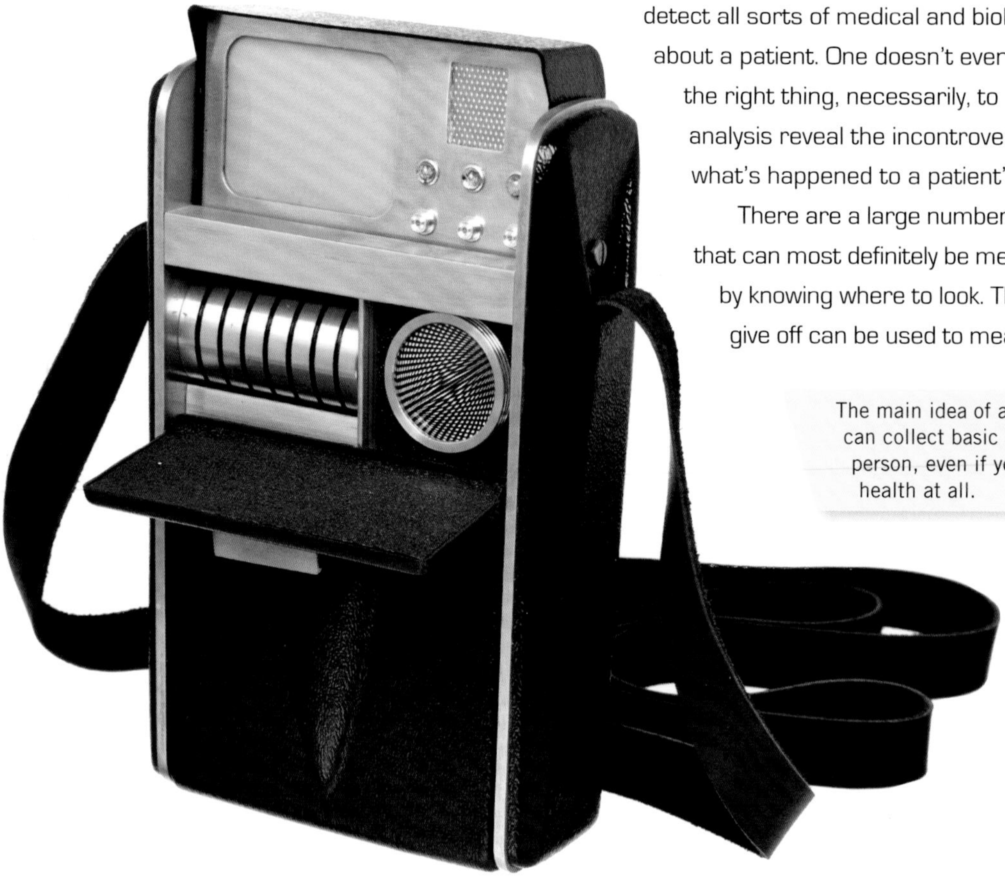

The main idea of a medical tricorder is that anyone can collect basic data about the health of another person, even if you have no training in medicine or health at all.

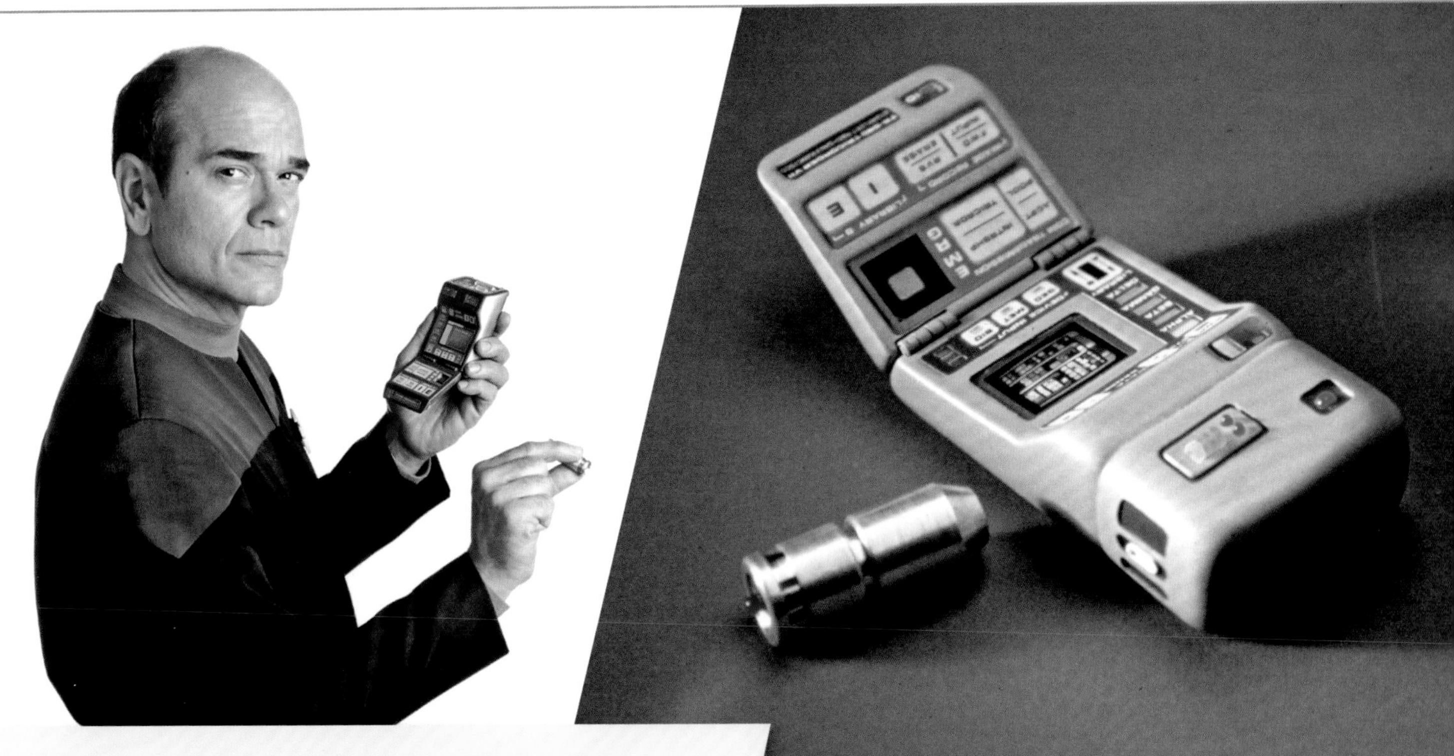

Tricorders could show any abnormalities in a patient's anatomy, physiology, or how their body was functioning. An indispensible forensic tool, it uncovered evidence that humanoids had been transporting through a dimensional shift, causing their tissues to degrade.

temperature with infrared measurements, visual and infrasound signatures can be used to track their pulse, and multispectral imaging can reveal subcutaneous injuries. In addition, certain samples—such as sweat, matter from the skin's outermost layer, and exhaled breath—can all be collected without contact. Finally, all the data can be analyzed by computer, a capability that's realistically been in place for more than a decade. By many metrics, then, the tricorder—which was a portmanteau of the idea of a *tri*function re*corder*, capable of sensing, computing, and recording all at once—has already become a reality.

Specialized devices, capable of monitoring conditions or testing for particular medical traits or levels, already exist in great abundance, such as the insulin and glucogon tracking, monitoring, and administration devices shown here.

On the other hand, people have been building progressively more advanced, self-contained devices in an attempt to bring the *Star Trek* tricorder to life. While numerous devices have been marketed as tricorders, the first serious contenders came forth in the 2000s. In 2007, a portable, self-contained, briefcase-sized mass spectrometer was introduced—a device capable of ionizing ambient chemicals without any additional preparation and sorting them by their charge-to-mass ratio. In 2008, a handheld, multispectral imaging device capable of detecting internal injuries regardless of lighting conditions or skin pigmentation was announced. In 2014, a device that purported to be a handheld DNA lab, capable of providing disease analysis and diagnosis for malaria in under fifteen minutes, claimed to be ready for clinical trials. And in 2016 and 2017, the Qualcomm Tricorder XPRIZE explicitly challenged teams to create a real-life tricorder: specifically, a device that could successfully diagnose over a dozen different medical conditions in a single, handheld device weighing less than 5 pounds.

Continuous health and bodily-function monitoring devices have taken off in recent years as well. The rise of wearable technology of all types, including devices like Fitbits and Apple Watches, keep track of the wearer's basic vital signs in real time, providing long- and short-term data to anyone who wants it. Single functions can be continuously monitored by a single device at present, but as miniaturization, computational power, and sensor capabilities all improve over time, it stands to reason that all of us, if we so wish it, might soon be able to equip ourselves with a single, wearable device capable of continuously monitoring our health. Temperature could be measured by an external thermometer or IR sensor; a microscope, smartphone, and a point of contact with the body (like a wristband) could monitor bodily secretions or epidermal bacterial content; electrodes could measure heart rate and the heartbeat's regularity; glucose levels can be measured via a tiny filament, eliminating the need for painful needle sticks; polarized light analysis can reveal wound healing or the presence/absence of cancer; portable DNA labs could monitor for the presence of abnormal mutations or the presence of specific antibodies; infectious germs could be monitored by external, airborne sensors; and ultrasonic probes can be used to take ultrasound images in conjunction with a smartphone. The individual components for a true medical tricorder all exist; we're simply waiting for them to be integrated into a single, powerful, available-to-all device.

As of the mid-2010s, there are a few devices in various stages of development aspiring to reach *Star Trek* levels of technology in a handheld, all-in-one device. Scanadu, a handheld device designed to be pressed against the user's temple, can read out temperature, heart rate, oximetry, ECG, respiratory rate, blood pressure, and emotional stress level all after just a few seconds of contact. With an additional (ahem) sample, it can perform a urine analysis as well. QuantuMDx Group, pioneers of the aforementioned handheld DNA lab, developed Q-POC, a virus-detection device that breaks open cells and analyzes their DNA, searching for viral signatures against a preprogrammed database. And Ibis Biosciences, which requires a blood sample, can search for signatures of over one thousand of the most common viruses, bacteria, and fungi that infect human beings and their cells. By comparing the pathogenic fingerprints of these invasive, single-celled creatures with an ever-growing database, rapid, accurate, and comprehensive early disease detection may be commonplace in the very near future.

Even in *Star Trek*, the tricorder was never designed to replace an entire sickbay, but rather to be a first line of inquiry in evaluating patient health. With the rapid rate of technological advances over the past twenty years, we've seen an incredible evolution of so-called "tricorders" that were no more than glorified barometer/thermometer combinations to handheld devices today that can perform a slew of medical information all at once, with new specialty devices capable of performing advanced diagnostics emerging all the time. If the current rate of technological advancement continues, it's reasonable to expect that a device that meets all the agreed-upon criteria of a tricorder, namely that it can:

- detect and diagnose disease
- continuously monitor health metrics such as heart rate, pulse, respiration, blood pressure, blood sugar, and more
- monitor overall health in both the short and long term
- summarize someone's state of health or illness
- and tell whether they're fighting or falling victim to an infectious agent

will not only exist, but will be widespread by the end of the 2020s. Not only that, but the smart bet is that it will be even smaller and more compact that the original *Star Trek*'s tricorder.

In the very near future, *Star Trek*–style tricorders will be available in a widespread fashion. Health and medical monitoring is already entering the era of big data and wearable devices, and the hope is that improved outcomes and truly personalized medicine will ensue. While some claim that the more data is available, the better off everyone will be, there are ethical issues surrounding self-treatment and attempts at self-diagnosis, as well as privacy and security needs that must be addressed. This was clearly still an issue as late as the twenty-fourth century, as tricorders were meant to be used by medical staff alone and were only obtained by hypochondriacs like Billy Telfer through theft. Despite the potential fears and issues associated with it, more data and more information almost always leads to better outcomes. Since longer, healthier lives are something almost all of us strive for, it's likely only a matter of time—and not that much time—before we all can look forward to a future where instead of rushing for a thermometer when we feel ill, we can simply reach for a tricorder.

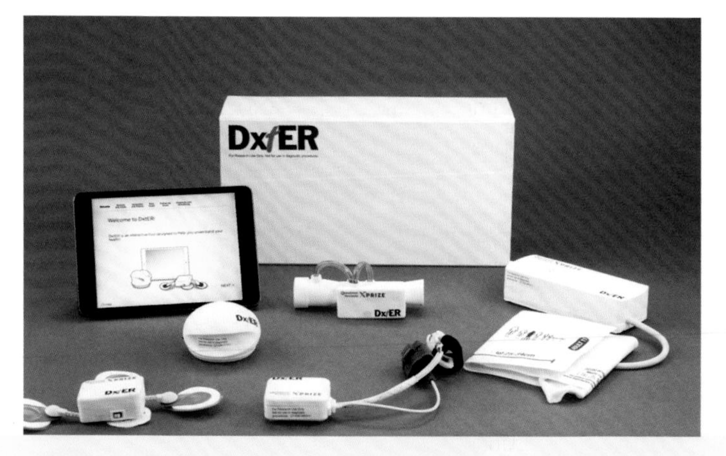

Final Frontier Medical Devices' DxtER prototype won the Qualcomm Tricorder XPRIZE competition in 2017—along with $2.5 million that will fund clinical trials.

CONCLUSION: MAKE IT SO

As scientists, we like to pretend that the way civilization advances forward is through discrete fundamental discoveries. That a new theory is developed, a new understanding is reached, and then applications come about. But *Star Trek*, perhaps better than any other cultural touchstone, demonstrates very clearly that it isn't the new discoveries that are always the driving force behind new advances for civilization. Sometimes, it's daring to dream. Sometimes, it's having the guts to look at the seemingly impossible—whether for practical or even fundamental reasons—and imagine how it might actually come to be. You likely won't be right down to the last detail, but having the vision of what the future could be is an integral part of what makes the human endeavor as worthwhile as it is.

Many of the technologies originally envisioned and brought into the public consciousness by *Star Trek* are so widespread now that we take them for granted. Transparent aluminum exists and is already being applied to everything from submarines to missiles. Real-time translators are coming into their own right now, both in spoken and written formats. Personal, tablet-style computers and phones with touchscreens are so ubiquitous that we think of life without them as being completely disconnected. We pass through sliding doors every time we walk into a supermarket and never think twice about it. Hyposprays were used for mass vaccination campaigns less than a decade after *Star Trek* first brought them onto our radar. And prosthetics, from robotic limbs to artificial eyes, aren't just on their way, but are already here. It didn't take centuries for many of these technological advances to come into being; it took less than a single human lifetime.

Other technologies look to be well on their way, as scientific advances have opened up tremendous new possibilities for us. Holograms are becoming more lifelike and can give tactile feedback and respond to their environment. Androids are becoming more and more humanlike with time, and chatbots get closer to passing Turing tests. (Although the passage of a Turing test may be more reflective of such a test's limitations than any true realization of artificial intelligence.) Computers can process natural language and speak back, crafting their own sentences to convey their point. 3D printers can replicate not only inanimate objects but some foods as well, while cloaking devices and tractor beams have had their proofs of concepts successfully demonstrated. We might not have achieved the dream of *Star Trek* just yet in every regard, but given that it's been just more than a half century since it first envisioned our world three hundred years into the future, our progress isn't too shabby.

We also mustn't forget to celebrate the various steps of progress, even if they haven't led us to the desired ends yet. Antimatter containment, deflector shields, cortical implants, and synthehol might not be in widespread use anytime soon, but we've learned a tremendous amount toward making them part of our reality. Antiparticles can be created, bound into neutral atoms, and precisely contained for long enough to have their properties—like atomic spectra—measured. Magnetically confined plasmas can be used to deflect away directed-energy strikes, one of the first steps necessary in the development of a deflector shield. Fundamental research in neurobiology is making the first rudimentary cortical implants possible, with a completely paralyzed human able to interface with a computer directly with her mind. And developments in the understanding of molecular binding in receptor sites and the chemicals they release has led to the development of many new pharmaceuticals, including along a path that may someday lead to synthehol.

Even technologies that may, in fact, violate the laws of physics are still worth exploring. Faster-than-light travel may be impossible given what we know today, but one doesn't need to violate Einstein's relativity to exceed the ultimate speed limit. Instead, controlling the folding of space may enable us to achieve that goal; it would only require the existence and harnessing of negative mass/energy. Inertia dampers may not be feasible if the equivalence principle holds, and gravitational and inertial masses are inseparable. But artificial gravity still may become real, albeit through other means. Subspace isn't a real thing, and so subspace communication is likely impossible. But combining the technologies of the folding of space with energy collimation—the latter of which is a presently emerging technology—may make faster-than-light communication a reality. And transporting a human being from one location to another by deconstructing and reconstructing their matter pattern may inevitably result in the destruction of the original, a fate we'd wish to avoid at any cost. But transporting the information encoded in an arbitrarily complex state is true in principle and becoming feasible in practice as we improve the technology associated with quantum teleportation.

Star Trek was inspiring as a television series and movie franchise, no doubt. But in terms of the technologies and the vision of the future it brought to humanity, it was truly transformative. It makes you wonder how many other technologies may someday come to fruition, all because humanity dared to dream them up.

But *Star Trek* also brought us a glorious future, where advances in science and technology were used for the benefit of all members of all planets in the Federation, by picturing scientists as altruists. Rather than warmongers who brought us progressively more destructive weapons, scientists were researchers who sought out truth, knowledge, and positive applications for peace and societal advancement. Sure, there might be phasers and photon torpedoes in the future, and there might be calamitous forces—hostile alien races like the Borg and the Jem'Hadar—that represented the worst of humanity's traits. But technology didn't just represent the power to bring ruin, but great power for creation and improvement as well. Shields could be extended to other ships; medical breakthroughs could save countless lives or even resurrect the newly dead; predictive modeling through computational simulations could even avert the most cataclysmic potential disasters. In the most trying of times, a civilization's greatest and most influential historical figures could even be brought back to life through cloning and memory implantation.

There are many problems facing the world today, both natural and stemming from humanity ourselves. Global warming and climate change are already changing the habitable landscape of our world, and the human activity responsible for the nonnatural component may already have passed the tipping point. Yet one of the countermeasures being seriously considered—geoengineering—was already implemented by the twenty-fourth century in *Star Trek* through the technology of weather control. The ability to control rainfall and cloud cover and to dissipate storms such as tornadoes and hurricanes before they ever cause harm would be an incredible boon to whatever planet it was implemented on. Before climate change was ever a serious threat, *Star Trek* was dreaming of ways that science and technology could shield the world from its most vicious influences.

But the biggest dream of all has yet to see any progress. Despite everything we've learned, both theoretically and experimentally, one phenomenon that appears numerous times in the *Star Trek* universe has eluded any semblance of progress: time travel. Time portals, transporter malfunctions, temporal vortexes, rifts, fluxes, or displacement, or simply a nearly omnipotent alien have all sent various characters or ships backward or forward in time, resulting in potential catastrophes to the timeline. Yet this appears to be one plot device that seems destined to remain in the realm of science fiction forever.

The only solutions to general relativity that admit time travel—requiring closed timelike curves—seem to be purely mathematical solutions, with grossly unphysical effects that don't appear to be feasible. Along with tachyons, chronitons, and verterons, some of our greatest hopes may simply hit an impenetrable wall when stacked up against the laws of nature.

Evidence for a potential technology's impossibility—practically, theoretically, or even physically—might be discouraging, but it mustn't stop us. Humanity, now more than ever, must continue to dream. The scientific discoveries that the past few decades have brought us have opened up possibilities that the original incarnations of *Star Trek* could never have envisioned. We now know our universe is filled with a mysterious, massive substance known as dark matter, outmassing the normal, atomic-based matter in our galaxy by a factor of five to one. It's expected that dark matter will turn out to be a new class of particle, with unique, yet-to-be-determined properties. It's conceivable that this could be a new energy source, which would be conveniently located anywhere we traveled within the Milky Way. Empty space itself, too, is filled with dark energy; the zero-point energy of the universe isn't zero, but rather has a positive, nonzero value. Perhaps at some point in the distant future, we could find a way to harness that and utilize it to further our own scientific and technological purposes and achievements.

Star Trek's vision for the future and how we can create a utopian society also showcased something rarely seen in modern fiction: entire civilizations working together for the betterment of all. When you look at many of the greatest advancements of the past few decades, they largely relied on humanity coming together on the largest scales—with funding and collaboration between the governments of many countries—to make it happen. The International Space Station, the Large Hadron Collider, the Internet, GPS networks, supercomputing, HIV treatments, the LED, and the Human Genome Project were only possible because of public investment in science and scientific research. The future of Earth that *Star Trek* envisioned required that our entire planet come together for the good of all of us, transcending national, racial, religious, and all other boundaries. In the human endeavor, as inhabitants of a pale blue dots in the abyss of a vast, expanding universe, we're all in this together.

Humanity has yet to begin our journey to the stars, but our journey into the future has brought us a safer, healthier, and more connected world than ever before. A smaller percentage of the world goes hungry, lives in poverty, or dies of preventable diseases than at any point in human history. Meanwhile, billions of computational devices—all more advanced than anything that existed at the time *Star Trek* premiered—have proliferated around the globe. The encyclopedic knowledge of everything known to humanity is available online over the Internet and accessible on computers, tablets, or smartphones from practically any location in the world. Our most fanciful dreams of what a technological utopia would be have come true in many ways, with many other aspects under development and destined to become reality in the near future. It is not in our nature to rest on our laurels and to merely enjoy and celebrate the achievements brought forth by our predecessors. Part of what makes us human is the journey to continuously strive for the next frontier, to cross the next boundary, to explore the newest unknowns, to unlock the next set of possibilities. Our journey does not end with us, but continues from generation to generation, with each one enjoying a better quality of life than the last. We have not reached our limits yet. Our mission to boldly go where no one has gone before continues. It's up to us to make it so.

Above all, *Star Trek*'s future has always been one of optimism and peace.

INDEX

ABOUT THE AUTHOR

Ethan Siegel is a PhD astrophysicist, science writer, author, (sometimes) professor of physics and astronomy, and longtime *Star Trek* fan. He has written for *Forbes*, *Scientific American*, NASA's *Space Place*, and many other print and online publications. His award-winning science blog, *Starts with a Bang*, has been educating the world since 2008.

IMAGE CREDITS

All images courtesy of CBS Studios Inc. except those on the pages noted below.

Alamy Stock Photo: 46 (Photo Researchers, Inc), 123 (ClassicStock), 134 (Aflo Co. Ltd.), 144 (Reuters), 171 (Reuters), 173 (ZUMA Press Inc), 181 (imageBROKER); **Bill Connelly:** 148; **CERN:** 49; **Corbis Historical/Getty Images:** 81; **Ethan Siegel:** 26 (based on work by Wikimedia Commons user Maschen), 47, 70 (based on work by Wikimedia Commons user V1adis1av), 80, 90, 112 (bottom), **Intel Free Press:** 206; **MIT BioInstrumentation Lab:** 202; **NASA:** 5, 36, 65 (Steele Hill/NASA /GSFC/SOHO/ESA), 93 (NASA/Laser Interferometer Space Antenna), 145, 161 (left), 163; **Nature Photonics**, 7, 123–127 (2013): 20 (Brzobohaty et al.); **Nature Physics**, 12, 355–360 (2016): 67 (M. Thévenet et al.); **National Institutes of Health (NIH):** 187; **Otis Historical Archives, National Museum of Health and Medicine:** 199 (photo by Roy C. McManus); **Pan American Health Organization:** 201; **Queen Mary University of London:** 58 (bottom); **Shutterstock:** 39 (Wojciech Tchorzewsk), 40 (HDesert), 48 (chromatos), 66 (MU YEE TING), 112 (top) (adziohiciek), 130 (cobalt88), 142 (OlegDoroshin), 149 (molekuul_be), 154 (BAHDANOVICH ALENA), 155 (RoongsaK), 188 (nobeastsofierce), 179 (Carolina K. Smith MD), 196 (Peter Hermes Furian), 197 (Alexilusmedical); **SpaceX:** 34; **U.S. Military:** 19 (U.S. Air Force), 41 (U.S. Navy/DoD photo by Petty Officer 3rd Class Brian Fleske), 56 (Defense Visual Information Center/U.S. Air Force), 116 (U.S. Army/Sgt. Austin Berner); **Werner Benger:** 91; **Wikimedia Commons:** 14 (AllenMcC); 58, top (Physicsch); 73 (Junglecat); **XPRIZE Foundation**: 208